Scrum for Hardware Design

**Supporting Material for
The Mechanical Design Process**

David G. Ullman
Professor Emeritus, Oregon State University

SCRUM FOR HARDWARE DESIGN

Published by David G. Ullman 621 Aeronca St. Independence Oregon 97351. Copyright 2019 by David G. Ullman. All rights reserved. No part of this publication may be reproduced or distributed in any form or by any means, or stored in a database or retrieval system, without the prior written consent of David G. Ullman, including, but not limited to, in any network or other electronic storage or transmission, or broadcast for distance learning.

ISBN 978-0-9993578-4-2

Ullman, David G., 1944- author.
    Scrum for Hardware Design / David G. Ullman, professor emeritus, Oregon State University.

ISBN 978-0-9993578-4-2

The Internet addresses listed in the text were accurate at the time of publication. The inclusion of a website does not indicate an endorsement by the author, and the author does not guarantee the accuracy of the information presented at these sites.

ABOUT THE AUTHOR

David G. Ullman is an active product designer who has taught, researched, and written about design for over thirty years. He is Emeritus Professor of Mechanical Design at Oregon State University. He has professionally designed fluid/thermal, control, and transportation systems. He has published over twenty papers focused on understanding the mechanical product design process and the development of tools to support it. He is founder of the American Society Mechanical Engineers (ASME)—Design Theory and Methodology Committee and is a Life Fellow in the ASME. He holds a Ph.D. in Mechanical Engineering from the Ohio State University.

More on Professor Ullman at www.davidullman.com

CONTENTS

Preface   vii

Scrum for Hardware Design   1

1. Introduction   1
2. Hardware Design Process   4
3. The Scrum Building Blocks   8
4. Organize   11
5. Plan the Project   20
6. Do - The Design Cycle   27
7. Review   31
8. Start Next Sprint Cycle   32
9. Mixing Scrum And Waterfall   33

Sources   37

Appendix A: A Glossary of Agile/Scrum Terms   37

Appendix B: Hardware Challenges   38

Agile Design of an Agile Fighter at Saab Aerospace   45

Introduction   45

Background   46

The Use of Scrum at Saab   48

Focus of the case study   49

Teams   52

Requirements and Stories in the Product Backlog   54

Tasks   55

Connected by Stable Fixed Interfaces   56

Retrospective   57

Conclusion   58

Acknowledgments   58

Sources   59

A Student Team Designs a Prosthetic Arm Using Scrum Methods   61

Introduction   61

Background   62

The Tools   64

The Process   66

What Was Learned   73

Acknowledgments   75

PREFACE

This book is a supplement to The Mechanical Design Process 6th edition (MDP6) adding material on the Scrum framework customized for hardware design. All mechanical design process texts including MDP6 are based on the waterfall, serial design methodology. While it is absolutely essential for hardware systems to be planned and be sequential from the customer needs, to concepts, to product, to manufacturing; the real-world forces much non-sequential effort that cannot be ignored.

In 2018 I began to develop material to integrate Agile methods into the text. Unfortunately, my knowledge about Agile was slim, at best. I had heard of Scrum as an Agile framework for software but knew little about it. I found that it was widely adopted in software development and, increasingly used in in hardware design to accommodate what cannot and should not be forced to be sequential. Because of its emphasis on managing uncertainty and team communication this method should be taught in mechanical and systems engineering education.

To develop the Scrum material included in this book I reached out to two people whose consulting practices focus on Scrum for hardware; Joe Justice of Scrum Inc. and Dr. Kevin Thompson of cPrime. They both trained me in the methods and graciously answered my many questions. They too think it important that Scrum become part of the hardware design curriculum.

One feature of *MDP6* is that it is supported by a second book of thirteen case studies (*The Mechanical Design Process Case Studies*). The case studies were developed over the last ten years with each co-authored by a practitioner applying some of the best practices featured in *MDP6*. These were originally only available on the web as individual downloads but put in softcover book form in 2018.

Scrum is new and different enough that I thought it important to include some case studies on its application. My search for assistance for this led to Professor Aaron Hoover and Jörgen Furuhjelm. Prof Hoover has taught Scrum as part of his mechatronics course and his work led to the first case study included here – "A Student Team Designs a Prosthetic Arm Using Scrum Methods." I found Jörgen through a web search as he has written many articles about the use of Scrum at Saab Aerospace. I later learned that he too was mentored by Joe Justice. Jörgen's work led to the second case study – "Agile Design of an Agile Fighter at Saab Aerospace."

Scrum for Hardware Design, the book you are reading, is also available in an ebook format. Both *MDP6* and its companion book of case studies are available only as paperbacks. It is not yet clear if the material here will become part of

MDP7 in paper or if *MDP7* will be put on the web. Like most design problems, the voice of the customer will drive future development.

ACKNOWLEDGEMENTS

When I began to write this material in Sept 2018, my knowledge of Scrum was very superficial. Thanks to Joe Justice, Jim Damato, and JJ Sutherland of Scrum Inc. I was able to take a five-day training course specifically targeted at hardware. Additionally, Dr. Kevin Thompson of cPrime Inc. worked me into a two-day course he was teaching at a customer's site, also hardware focused. These two hands-on courses got me up-to-speed on Scrum.

The material Joe and Kevin teach is ideal for industry but not fully suited for college engineering courses. To help make this material better fit an academic environment I conducted a survey of professors interested in design education. This survey not only probed interest in Scrum for hardware where 87% of the responses said they were interested in learning more, I also asked if anyone knew of Scrum for hardware courses currently being taught.

One of the responses was from Professor Aaron Hoover at Olin College. He has graciously let me use material from his mechatronics course in the body of the book and as one of the two case studies. His approach is to have the students use Scrum to design a mechatronic device. In this manner he teaches a technical topic and a design process at the same time. He has worked closely with Professor Lawrence Neeley also of Olin who has also worked with Professor Charles Kim of Bucknell. All three of these professors kindly supplied me with material on how to teach Scrum in a university environment.

All this material has been developed with cooperation with my colleague Joshua Tarbutton of the University of North Carolina Charlotte. As this work was being published, he was applying it to a senior project class in order to test it.

Scrum for Hardware Design

KEY QUESTIONS
- What are agile and Scrum? What are the steps to do Scrum?
- Why is the Scrum method important to hardware design? How does a Scrum sprint result in a well-designed product?
- How are hardware and software design different, and how do these differences affect the Scrum method?
- How does the Scrum method fit with the waterfall method?

1. INTRODUCTION

One of the most challenging problems faced by designers is the necessity of making decisions when still unsure of what a good solution will look like and how it will function. This uncertainty characterizes the design process. Since design is learning, when you know enough to design a product you have already created it - a design process paradox[1]. This paradox is true of virtually all design projects be they for software or hardware. Traditional design processes are often too rigid, limiting creativity and forcing projects over budget and late to market.

"Agile" methods have recently been developed to respond to the design paradox. Agile means what it says - the ability to respond effectively during the design process to uncertain, risky, variable, and evolving information. Both the product and the process need to be agile in these environments.

Agile methods were developed for software design beginning at the turn of the 21st century. They are now a predominant method for developing code. Hardware and systems design have been slow to follow. There are many reasons for this lag (See Appendix B for details); however, farsighted companies like

---

[1] See Fig. 1.9 in *The Mechanical Design Process*, 6th edition for a graphical representation of the design paradox.

John Deere, Saab Aerospace, SpaceX and Tesla now apply agile methods across all disciplines. This chapter describes how to do just that.

Agile product development is based on an incremental, iterative approach. Instead of in-depth planning at the beginning of the project like the sequential waterfall (aka stage-gate) method, the agile process is open to changing requirements over time and encourages constant feedback from the customers. The differences and synergies between the sequential and incremental methods are covered in Section 9.

While there are many agile methods, the Scrum framework, initially developed for software design, is the most widely used and has the highest potential for impact on the mechanical design process. The term "Scrum" was taken from Rugby. In rugby, the ball gets passed from team member to team member as it moves as a unit down the field, so a Scrum is a team of people interacting together for the greater good. In the case of Rugby, moving the ball is the goal. For design, developing the best product (in the eyes of the customer) in the least time possible is the objective.

The goal in this chapter is to introduce the Scrum framework for agile hardware design ("hardware" being everything but "software"). Scrum provides a mechanism for planning and tracking work, leaving the detailed nature of the design work up to the team. There are many different styles of Scrum, so the material in this chapter may not fully fit how it is applied in all companies.

1.1 Agile's Principles

Agile methods, and thus the Scrum process, is based on ten principles. These ten are modified from the original set of twelve developed for software to make them more general.

1. **Build customer satisfaction through early and continuous delivery**: Customers are happier when they receive working prototypes, simulation results or products at regular intervals, rather than waiting till the end to see the final result as is common in the waterfall process. Scrum accommodates this principle with the design team operating in short sprints or iterations that ensure regular progress.

2. **Accommodate changing requirements throughout the development process**: Traditional processes emphasize developing the engineering specifications at the beginning to design process. But, only for very mature products can you know all the requirements from the beginning. Most projects have many others evolving as the product matures. Agile methods avoid delays when requirements change, or a new feature is added. Note that this principle says, "accommodate." There is a balance between accommodating change and "fostering change". Fostering change encourages "feature creep" when new features are added to a product or changed at a pace that makes it impossible to finish the design work. Traditional methods fight feature creep by forcing up-front specification development and then making them hard

to change; agile does not. Thus, this principle is a feature, but care must be taken.

3. **Build collaboration between the business stakeholders and developers throughout the project:** Better decisions are made when the business and technical functions in the organization are aligned. Alignment requires frequent communication. In many companies, the business units seldom talk with the product development units. Often the interaction is limited to an initial meeting to define the product and at product reviews when the product is nearly completed. Agile methods force much tighter interaction.
4. **Build projects around motivated individuals.** Motivated teams are more likely to deliver their best work than are uninspired teams. In the Scrum methodology, small teams self-govern, make decisions and are supported to get the work done. This trust in the team builds motivation.
5. **Enable face-to-face interactions** – Communication is more successful when development teams are co-located. As will be seen, Scrum methods encourage development teams to do all work in the same office to encourage collaboration.
6. **Support sustainable development and a consistent development pace:** Teams establish a repeatable and maintainable speed at which they can deliver results, and they repeat it with each design cycle.
7. **Rely on technical excellence and good design to enhance agility**: The right skills and good design processes ensure the team can maintain the pace, constantly improve the product, and sustain change.
8. **Design for simplicity:** Scrum teams develop just enough to get the job done for the current needs.
9. **Encourage great architectures, requirements, and designs.** The structure of Scrum forces early attention to product architecture and the evolution of requirements and product.
10. **Reflect on how to become more effective**: Scrum has built-in self-reflection leading to process improvement, and advancing skills and techniques help team members work more efficiently.

The rest of this chapter will show how to realize the ten principles during hardware design using the Scrum framework. Agile is more than a project management method; it is a state of mind. The Scrum methodology presented here helps internalize the ten principles and build better design habits.

A warning!! "Agile" speaks of quickly reacting, not rushing. However, doing design fast does not mean paying less attention to detail. Product quality is designed in, and just because the Scrum method breaks design into short, fast sprints does not mean that any less work can be done to achieve a quality product.

1.2 An Example

To show how Scrum supports the ten Agile Principles, a running example flows through this chapter. This is an abbreviated version of a more complete case study. This study and another focused on the design of Saab's Gripen JAS 39 appear later in this book.

A multi-disciplinary team of four students at Olin College embarked on an eight-week project to design a below-the-elbow prosthetic for amputees. The prosthetic hand needed to sense when to grip and when to release based on body movement and tactile feedback. The gripper in Fig. 1 is part of the arm designed by the Olin team.

Generally, student teams at Olin are multi-disciplinary and here: Victoria was focused on sensors and software; Celine also had sensor knowledge; Liani was interested in mechanical systems and manufacturing; and Ellie, a mechanical and manufacturing engineer, was most interested in human interfaces. An added self-imposed challenge for the team was to design the feedback control system using only readily available do-it-yourself technology. The students were all sophomores except for Victoria, a junior.

Figure 1 The final gripper on the Olin prosthetic arm

2. HARDWARE DESIGN PROCESS

This section describes the basics of the Scrum process for hardware design. To best understand how it differs from other processes, consider the three variables that can be controlled and traded off against each other during design: scope of work, budget, and time.

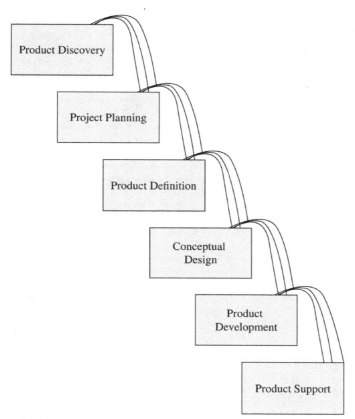

Figure 2 The Waterfall Process - Reprinted from figure 5.13 in *The Mechanical Design Process, 6th Edition*.

In the traditional waterfall process (Fig. 2), the scope of work is fixed (the product requirements set at the beginning), the cost of the product predefined, and time managed with Gantt charts. This all works if the product and associated processes are well known, and uncertainty is low. However, often the lack of knowledge and unexpected events cause the project to fall behind and creep over budget - the schedule is relaxed as are the cost targets.

Fig. 3 shows a typical Gantt Chart used for planning. It itemizes the tasks with estimates for start and end dates and progress tracked by changing the color of the bars.

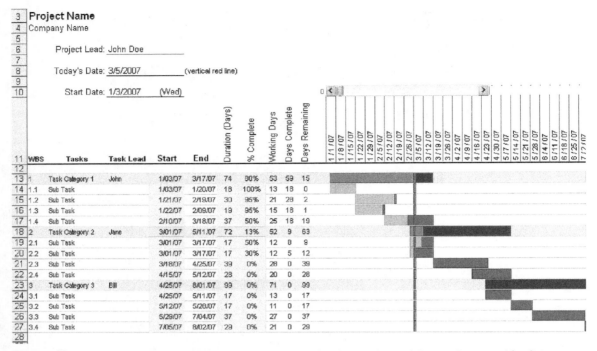

Figure 3 A typical Gantt chart, reprinted from Fig. 5.22 in *The Mechanical Design Process, 6th edition*.

Limitations of waterfall and Gantt charts are:
1. They are not often updated. They are built at the beginning of a project and then, for the most part, unchanged. Since tasks change and dates slip, a Gantt chart is always out of date.
2. If too detailed then a Gantt chart can be hard to read and even more likely to be out of date as the management overhead is just too high to keep it current.
3. Gantt charts are typically associated with a waterfall process. While work on multiple tasks can be seen in Fig. 3 (the vertical bar – the current state – shows five tasks in process), the actual progress toward getting the project done can be lost, especially if the chart is not very detailed.

In Scrum, time is fixed and broken into short "sprints," and scope of work flexible. This implies that the design team has some discretion over exactly what features make it into the final product and how they are implemented. In fact, the Scrum method is "timeboxed" with all the activities constrained in time. While this may seem impractical, it works quite well when the requirements, technologies or the market are uncertain.

Waterfall is best for known systems and Scrum best for when uncertainty is a factor. The interaction of the two is saved for Section 9. For now, the basics of Scrum will be developed.

The Scrum method uses 2 to 4-week time-bound design efforts called "sprints." Saab's Aerospace group uses 3-week sprints, and Volvo Trucks' sprints are 2-weeks long. During each sprint, Fig. 4, the team <u>organizes</u> and defines their goals for the product, they <u>plan</u> how to achieve the goals, they <u>do</u> the work, <u>deliver</u> the product and then <u>review</u> their work to find out how to improve both the product and process during future sprints.

Where the larger cycle - the Sprint Cycle - is focused on the design process, the Design Cycle - the "Do" portion of Scrum structure - is where the actual design work is done. During planning, early in the sprint, the team decides what will be designed and who will work on what. The design work is self-managed and is drawn as a smaller circle as it has a faster, daily pace.

When the team is finished with one sprint, they tackle the next one and so on as shown in Figure 5. At Saab Aerospace, the sprints are end-to-end, beginning and ending every three weeks. If the sequence of sprints is all focused on the same function, then they are simply iterations, each sprint refining what was accomplished in the previous sprints. During the first sprint the team may develop concepts, and in the second choose the best concepts, develop prototypes in the third, and so on.

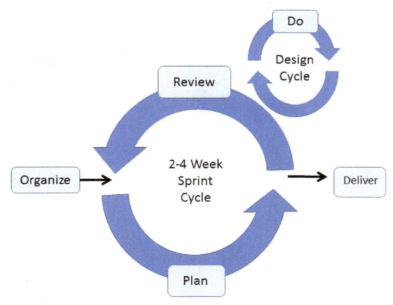

Figure 4 The two Scrum cycles in a sprint

Figure 5   Scrum sprints in a series

Working in a sequence of sprints becomes routine with some of Saab's mature Scrum teams working this way for over ten years. The approach changes the emphasis from meeting performance (needs) goals as represented on Gantt chant to generating small, incremental results as fast as reasonably possible. This structure is the framework for fulfilling the ten Agile Principles.

3. THE SCRUM BUILDING BLOCKS

To do Scrum - to execute the sprints - there is a process to be learned. There are thirteen process objectives as summarized in Table 1 and shown in a pictograph in Fig. 6. Material in this chapter details each of the steps. The Scrum vocabulary is introduced as needed and itemized as a glossary in Appendix A.

Both the table and figure are built around the same Organize-Plan-Do-Review structure as in the previous figures. Also shown are the steps (the activities that describe what is to be accomplished); the artifacts (the result of the activity) and the meetings to be held.

3. The Scrum Building Blocks

Table 1 The Steps of Scrum

		Process Objective	Activity	Artifacts	Meeting
Organize	1	Organize Team	Choose team members and identify the Product Owner (PO) and Scrum Master (SM).	Team Roles	
	2	Develop Product Goals	Generate Product Goals in terms of user stories or requirements.	Product Goals	
	3	Create Scrum Board	Build Scrum Board (either physical or in software) with areas for Product Backlog, Sprint Backlog (To Do), Doing and Done.	Scrum Board	
Plan	4	Rank Order Stories	Rank order stories based on dependency, uncertainty and importance.	Product Backlog	Product Backlog Grooming
	5	Choose Stories for Sprint	Choose stories to be addressed in the Sprint.	Sprint Backlog	
	6	Identify Tasks	Identify the tasks that need to be done to meet Product Goals complete with measures, targets and tests that define when they are done (i.e. test-driven development).	Tasks for Sprint Backlog and Product Backlog	Sprint Planning Meeting
	7	Estimate Task Time	Estimate the time each task will take.		
	8	Choose Sprint Tasks	Choose what tasks to complete during the Sprint. Only start what you can finish.		
	9	Align Team Members with Tasks	Choose which team members are responsible for which tasks. Move tasks in-process to "Doing" area of Scrum Board.		
Do	10	Do Work	Do the technical work moving tasks from "To Do" to "Doing" to "Done" on Scrum Board.	Product Artifacts	Sprint Standup
	11	Track Progress	Track progress on Scrum Board Burndown Chart.	Update Scrum Board	
Review	12	Review Sprint Product	Hold a Sprint Review (aka design review) where the progress on the product is demonstrated.	Updated Product Intent and Tasks	Sprint Review
	13	Review Design Process	Have a Sprint Retrospective where the design process is reviewed, and improvements developed.	Team Process Changes	Sprint Retrospective

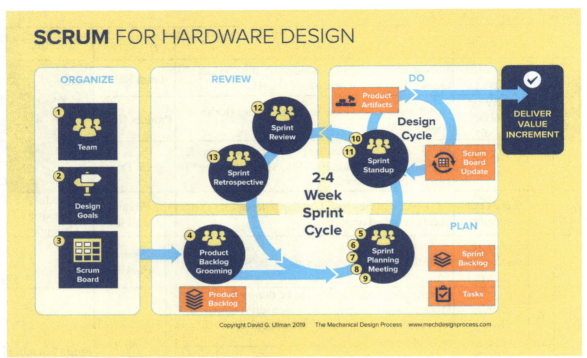

Figure 6 The Detailed Scrum Process

In Fig. 6 the meetings that manage the process and the artifacts produced are shown superimposed on the two cycles. The numbers relate to the sprint steps itemized in Table 1 and guide the sections below.

During each sprint there are six types of meetings[2] as shown. Typically, these meetings take 10%-15% of each team member's time. While this may sound excessive, the meetings are structured so that they actually save time as they keep communication open, brief and on-point.

To give some sense of what is to come, the process begins by organizing the team (1), defining the goal for the project – the goals to be achieved reflected through the voice of the customer (2), and establishing a Scrum Board (3) – a structure to manage information much like a Gantt chart in the waterfall process, but very different. The goals are organized and ranked forming a Product Backlog (4) during Product Backlog Grooming meeting. The goals to be addressed during the next sprint are chosen to form the Sprint Backlog (5). These are posted on the Scrum Board. For each 2-4 week sprint cycle, there is a Sprint Planning Meeting where the team identifies the actual tasks that need to be done during the sprint (6), (7) and (8) and who will do each of them (9). The tasks are added to the

---

2 In Scrum jargon, "meetings" are often called "ceremonies."

Sprint Backlog on the Scrum Board. During the sprint, the team self-organizes to do the tasks on the Sprint Backlog.

Next comes the Design Cycle which very similar to any other design process only broken into shorter chunks. The tasks in the Sprint Backlog drive the technical work as concepts are conceived, analyzed, built and tested. This work results in Product Artifacts (10) - the output of the sprint. The completion of a task is updated on the Scrum Board (11). This cycle repeats during the sprint with effort guided by the tasks in the Sprint Backlog.

One difference from traditional design work is the Sprint Standup meeting. This is a daily 15-minute meeting where the team shares the status of their work.

At the end of the 2-4 week sprint cycle there are two additional meetings aimed at improving the future sprints, The Sprint Review (12) is when the team evaluates and demonstrates the technical results of the sprint and, second, the Sprint Retrospective (13) when the team reviews how they managed the process and identifies how they can improve on it during the next sprint.

While this may all seem very complicated, it is not. Scrum provides a structure that enables the design team to do the best work they possibly can in the least amount of time. It establishes an environment that can motivate individuals, support their needs, and shows trust in them to get the job done. The following sections detail the 13 steps one-at-a-time, so they can be applied, the jargon learned, and quality products designed. To help with the new terms, there is a glossary in Appendix A.

4. ORGANIZE

As with any process, organizing at the beginning is necessary for success. In this section, three important project aspects are covered; forming the Scrum team, defining the product goals and setting up a Scrum Board.

4.1 Create Scrum Teams

	Process Objective	Activity	Artifacts	Meeting
1	Organize Team	Choose team members and identify the Product Owner (PO) and Scrum Master (SM).	Team Roles	

Scrum Teams have 4-9 members who carry out all the design tasks: plan, conceive, analyze, build, test, decide, and document; and they self-manage the process. There are three roles on the team: The Product Owner is responsible for what the team works on channeling the voice of the customer; the Scrum Master is responsible for making sure the team can operate how it needs to; and the Technical Team Members do the design work designing deliverables that respond to the voice of the customer. This team structure allows decision making at the lowest possible level in the organization. It makes the bureaucracy smaller, makes decisions faster and empowers the engineers.

The Product Owner (PO) is the voice of the customer. The PO is the team's face to the outside world. They define the specifications and decide on the sequence of deliverables. To fill this role, the PO must be:
- Knowledgeable about the domain. This is not to say that the PO must be an expert in any specific product function or have knowledge about all the functions but must have a general understanding about the technologies that might be used during its design and manufacture. Additionally, they must be knowledgeable about the market for the final product as they are the conduit between the team and the voice of the customer. The PO is often a marketing person.
- Empowered to make decisions.
- Available to the team to explain what needs to be done and why. Accountable for product value.

The Scrum Master (SM) drives the process and is accountable for removing impediments, so the team can efficiently complete the design cycle. Of all the roles, the SM is the most out-of-the-ordinary. They:
- Enable the process. They must be knowledgeable about Scrum methods and see to that they are followed. The Scrum Master does not have any say on the actual design process that the team follows in their day-to-day activities, only that whatever they do must adhere to the flow described in this chapter.
- Facilitate meetings.
- Maintain situational awareness of the work.
- Remove obstacles and protects the team from interference.
- Manage people and processes. The Scrum Master is the closest role to a manager in the traditional sense. Where traditional managers would try to control the team, the SM focuses on protecting the team from interruptions and removing obstacles for them.

The remainder of the team members is the "Technical Team" – the doers. The Technical Team should consist of 2-7 members and have sufficient technical experience to do the work. They are engineers, designers, analysists, testers, and others who bring expertise to the team.

In total, the Scrum team = PO + SM + Technical Team. What is unique about Scrum teams is that they are self- organizing. In-other-words, they decide what to do and how to do it to meet the needs communicated through the Product Owner. Other important features of Scrum teams are:
- Everyone is on one and only one team. Within Saab Aerospace almost all engineers work only on one team.
- Teams are stable; some are together for years.

- Teams are not interrupted during a sprint. The Scrum Master protects them from being bothered. Teams do not work much overtime.
- Teams are productive.

Sometimes there is not a sufficient number of people to have dedicated PO or SM on the Scrum team. In these cases, one person can fill multiple roles. However, the SM and PO should not be the same person as too much authority is then held by one person. The SM can be on the Technical Team, but this is not optimal as they must both focus on the technical work and be aware of the whole process at the same time, which may not be possible. The PO can be on the Technical Team but is usually a marketing focused individual and may not have the technical expertise.

On the Olin College team designing the prosthetic arm, the course professor served as the Product Owner and he also served as Scrum Master assisting the team through the process. At Saab, there are specific PO and SM team members with one PO serving multiple teams.

The ideal Scrum team, including the PO and SM, should be collocated, all in one room. Collocation ensures the most efficient and effective method of sharing information within any group of people, face-to-face. Ideally, the entire team has desks in this room, prototypes of the systems they are working on are here, and there are sufficient tools, so they make changes to the prototype as they work. If split up geographically, then collaboration suffers.

4.2 Develop Product Goals

	Process Objective	Activity	Artifacts	Meeting
2	Develop Product Goals	Generate Product Goals in terms of user stories or requirements.	Product Goals	

Developing goals is an essential step because, if you don't know where you are going, you won't know when you get there. A product can only be as good as your understanding of the needs of its' customers and fulfilling these needs is the goal.

For hardware systems, the product goals come from two overlapping sources: requirements and stories. In the traditional world of product development, there is a significant effort at the beginning of the project to identify the product requirements. Methods to support this are detailed in Chapter 6 of *The Mechanical Design Process* and built around the Quality Function Deployment method, aka the House of Quality. This up-front effort is contrary to the agile philosophy where the needs coevolve with the product itself using Customer Stories. Customer Stories are unique to agile methods, and as with the QFD, aimed at translating the voice of the customer into clear and testable requirements for what needs to be designed.

Each of these two methods has its strengths and weaknesses as shown in Table 2. Best practice for hardware design makes use of both methods. Organizations like Saab Aerospace mix the two, working out the high-level requirements at the beginning of the project using a form of QFD, and allowing these to drive the development of more detailed goals during each sprint.

Since the QFD is well developed in *The Mechanical Design Process*, only customer stories are detailed here. To be clear, these two methods significantly overlap. A good effort at traditional specification development includes all the customer stories, and a good job developing customer stories uncovers all the critical requirements. The addition here is that QFDs tend to be static documents, whereas customer stories continue to evolve during the project as will be seen.

	Requirements definition using QFD	Customer Stories
When in design process best applied	Beginning	On going
Detailed method to understand need	Excellent	Good
Responsive to unknowns	Forces study	Excellent
Guards against feature creep	Good	Poor
Agile relative to uncertainty	Poor	Good
Guides understanding relative importance	Good	Poor
Helps understand dependences	Good	Poor

Table 2 Comparison between QFD and Stories

Stories are Scrum's way to capture the requirements in the voice of the customer. The best format for a story is to complete the following statement:

"As a < (customer role or system)> I want to <perform an action> so that I can <gain this benefit>."

Alternatively,

"As a <who>, I <want what> so that <why>."

In these sentences the "action" or the "what" refers to the desired function of the product. The "gain this benefit" or the "why" is the requirement that the product needs to meet.

Stories are written primarily by the Product Owner with the help of the Technical Team and with input from the users or customers for the product. Capturing customer stories serves the same function as the first two rooms in the QFD which emphasize identifying the customers and what they want in the product. Stories have a more rigid format that helps capture the needed information.

It is handy to write each story or requirement on a sticky note like that shown in Fig. 7. This way the stories can be stuck on a wall and easily rearranged as will be explained in the next step.

Story or Requirement Title		ID	
Narrative			
Acceptance criteria			

Figure 7 Story or Requirement Card

To help with bookkeeping also give each story an ID number. It makes no difference if they are sequential or even continuous, as long as each has a unique number.

The narrative begins with a story in the format described above ("As a <role> I want …), followed by anything that supports the story: text, links, lists or drawings.

The bottom field, the acceptance criteria, are the measures used to determine when the story is done. These should describe tests to do or some feature that will be measured. The concept of generating good acceptance criteria is further developed in Step 6 when defining the sprint's tasks.

A useful mnemonic to use when writing the narrative and acceptance criteria is that good stories INVEST; they are:[1]

- Independent. The story must be actionable and "completable" on its own. It shouldn't be inherently dependent on another story, although this is not always possible. If a story is too complex, break it into manageable sub-

stories[3]. Independent stories are often difficult to write for hardware, so the dependence on other stories should be made clear from the beginning or as soon as they become evident.
- <u>N</u>egotiable. Until a story is being done, it needs to be able to be rewritten. Scrum activities have the allowance for changes built into the process. Sometimes the details of a story can't be known until its partially completed.
- <u>V</u>aluable. It delivers value to a customer or stakeholder. Some stories are more inward focused, not customer facing, and they are valuable to some other system.
- <u>E</u>stimable. The team must be able to estimate how long it will take to complete.
- <u>S</u>mall. The story needs to be small enough to be able to estimate easily. If it is too big, rewrite it or break it down into smaller sub-stories.
- <u>T</u>estable. The story must have a test it needs to pass to prove it was completed. Write the test before the team does the story.

One goal in writing stories is that the PO and Technical Team should capture stories about everything that could be done on the project, ever.

The team from Olin College designing the prosthetic used a web-based project management application that allowed them to manage "cards" (see Fig. 11), each card representing a story or task (see Steps 4-8).

Their two top-level stories were:
- As an amputee, I want an inexpensive prosthetic arm that can grip and release using an intuitive interface for control so that I can gain mobility at an affordable price.
- As a team member, I want to design a prosthetic device using off-the-shelf DIY sensors and controllers so that I can learn about these technologies.

They rapidly broke the first story into many sub-stories which were written more as technical requirements:
- The hand should be able to grip soft objects without crushing them.
- The feedback cuff should provide pressure proportional to grip force. The prosthetic arm should fit most users.

All the examples above lack "acceptance criteria" and thus are relatively weak. An excellent feature of the QFD method is that is forces testable criteria to be developed along with the stories.

An example in Chapter 6 of *The Mechanical Design Process* shows a small part of the QFD for the design of an aisle chair - a wheelchair used by airlines to help people move from their personal wheelchair to their seat on an aircraft. It must fit down the aisle of the plane, hence its name. In that example, one set

---

3 In Scrum jargon, a large story is an "epic" that is broken down into some smaller, manageable stories.

of customer requirements looks as shown on the left in Fig. 8. The top-level requirement is equivalent to a super-story with the other requirements being sub-stories. The highlighted requirements are rewritten in story format on the right. In the full QFD, there are many super-stories and sub-stories.

Sample from QFD			In Story format
Transfer from personal to aisle chair	Aisle chair preparation	Easy positioning of seat height	1. As a wheel chair user I would like to able to easily transfer from my personal wheelchair to the aisle chair to make it easy for me and the airline to accommodate me
		Easy to position chairs	1.1 As an airline gate agent I would like to be able to prepare the aisle chair to make the transfer safe and easy.
	Passenger movement	Minimum effort for all	1.1.1 As an airline agent I would like to be able to adjust the seat height easily to make a safe transfer of the customer from their personal chair to the aisle chair.
		Good lifting position	1.1.2 .....
		Minimum time for transfer	1.2 ..... 2. ......

Figure 8 The equivalence of user requirements in the QFD to user stories

For simplicity, from here on all stories and requirements will just be called "stories." This is consistent with Scrum terminology and is easiest to read.

4.3 Create Scrum Board

	Process Objective	Activity	Artifacts	Meeting
3	Create Scrum Board	Build Scrum Board (either physical or in software) with areas for Product Backlog, Sprint backlog (To Do), Doing and Done.	Scrum Board	

The Scrum Board is the primary tool used by the team to manage the sprint. There are many variations, but the ones shown in Figs. 9, 10 and 11 are typical and have all the essential areas on it. The use of this board is a form of visual planning and communication that makes the Scrum process very powerful.

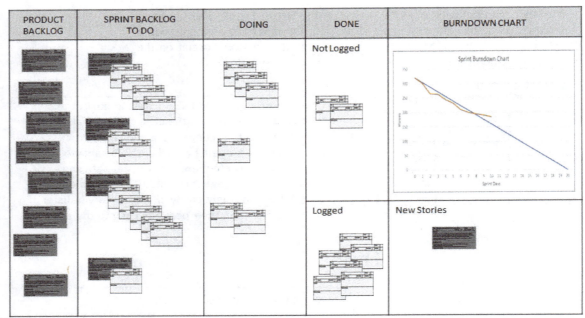

Figure 9 A Scrum Board

The Scrum Board can be a wall with tape defining the different areas or a white-wall that you can draw lines on as in Figs. 9 and 10; or can be on a web-based project management system as in Fig 11. The stories (dark icons) on Fig. 9 are on sticky notes and can be moved from area to area as seen on an actual Scrum Board in Fig. 10.

Beginning on the left, the PRODUCT BACKLOG begins with the stories captured in the previous step. Step 4 refines this area. The SPRINT BACKLOG has the stories and their supporting tasks that will be addressed (i.e., TO DO) in the current sprint – described in Steps 5-9. From the tasks in the Sprint Backlog, the team is currently working on those in the DOING section – Step 10. The team members chose these during the Sprint Planning Meeting as the tasks to address.

4. Organize

Figure 10 Scrum Board with tasks on yellow sticky notes.

Completed work on the tasks is moved to the DONE area, and progress tracked (Step 11) on a BURNDOWN CHART.

The Olin team of students managed their tasks online, Fig. 11[4], using a commercially available software called Trello. There are many other similar project organization systems. With systems like the one they used, the virtual cards can easily be moved from left to right as they are completed, edited, or new ones added.

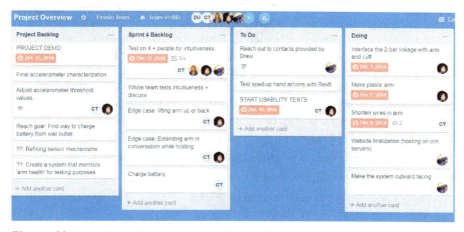

Figure 11. Part of the Olin team's online Scrum Board

---

4 The Olin team used Trello, but there are many other task management systems available.

5. PLAN THE PROJECT

With the team, the intent and Scrum Board in place, the Scrum process can get started. The Product Owner manages this up-front work with the support of the Technical Team in a Product Backlog Grooming meeting. The goal of this meeting is to choose which stories to address in the sprint.

5.1 Rank Order Product Goals

	Process Objective	Activity	Artifacts	Meeting
4	Rank Order Stories	Rank order stories based on dependency, uncertainty, and importance.	Product Backlog	Product Backlog Grooming

The Product Backlog area on the Scrum Board is the sum of all the stories that the team has identified. Now the team must choose which ones to work on during the next sprint. This is the first planning step and is the goal of the first sprint meeting, Product Backlog Grooming. This type of meeting is held at the beginning of the sprint but is also held weekly to update the project's goals. While the weekly meetings are limited to one hour, the initial meeting might take longer.

Choosing which stories to tackle in the sprint requires several sub-steps. First, the stories must be rank-ordered. Then the amount of time needed to finish the most important stories estimated. Finally, the team chooses the stories to complete during the sprint. The stories not included in the current sprint remain in the Product Backlog to be reconsidered for some future sprint.

The team rearranges stories so those with the highest value and highest risk or uncertainty are on top as shown in Figs 9 -11. Exact methods for ranking are up to the team[ii]. When ordering pay particular attention to:

1. Dependency: What stories are dependent on it and what is it dependent on?
2. Uncertainty: How certain is the knowledge needed to meet the story?
3. Importance: How important is it to meet the story?
4. Lead Time: Does the goal need to be addressed early in the process because there are outside factors that require a lead time?

The best approach is to ask each story, "Is completing this story vital to the project?" or "How critical is it to resolve the uncertainty in this story?" Optionally, compare two stories and ask, "Which story is most critical to the project and has the highest uncertainty?"

If a story seems too complicated and there are parts of it that can be done early, and other parts that can be done later, then decompose it into sub-stories.

The ranking developed here may be revisited during future Backlog Grooming meetings.

The Olin team reduced their stories into sub-stories. In an abbreviated form, some of those identified were:
- The arm needs to be able to sense commands to grip and release.
- The arm needs to grip odd shapes
- The team needs better understand microcontrollers
- The prosthetic needs to fit amputees' arms.

Finally, for the topmost stories, estimate the time needed to complete each of them in person-hours. This type of estimate will probably be very inaccurate, but it is only needed to get the process going. It will be refined in Step 7.

5.2 Choose Stories for the Sprint

The work to be done during the sprint is chosen here, during the Sprint Planning Meeting, and refined in Step 8, Choose Sprint Tasks. Since the Scrum process is timeboxed, fixed at a given length, the goal is for the team to choose what they think they can get done during the time available. As already said, the estimated time to complete a story may be very crude at this time and will be refined later.

	Process Objective	Activity	Artifacts	Meeting
5	Choose the Stories for the Sprint	Choose the Stories to be addressed in the Sprint.	Sprint Backlog	Sprint Planning Meeting

Assuming a sprint is 2-4 weeks long and there are N Technical Team members, then the number of stories that can be completed in the sprint are easily calculated. Say for example there are 4 persons on the Technical Team and that the sprint is 3 weeks long, then the total number of Technical Team hours are:

Total hours = 3 people x 8 hours a day x 5 days a week x 3 weeks = 360 hours

This estimate needs correction for the fact that, 10% - 15% of the time goes to meetings and there may be sick or vacation days during the time. Thus, a better estimate might be 270 hours. So now the question is, how many of the stories can the team complete during the sprint?

Typically, a team of eight people (PO, SM, and six Technical Team members) can complete 5-15 stories in a 3-week sprint giving an average of about 100 person-hours per story. These 100 hours may be spread across three people working for a week, one person working for 3 weeks or any other combination the team chooses.

For example, say that the ranked list of stories looks as shown in Fig. 12:

Story Rank	Hours to complete	Cumulative hours
1	50	-
2	120	170
3	80	250
4	100	350
5	10	390

Figure 12 Choosing the sprint's stories

In our example of 270 sprint hours, the Technical Team should be able to complete the first three stories (250 hours), but they can't finish the 4th story. Stories that can't be finished should not be started. So, the team should commit to doing stories 1, 2, 3, and 5 for a total of 260 hours. Story 4 remains in the Product Backlog with other stories and considered for the next or some future sprint.

All the stories to be completed during the Scrum are entered in the Sprint Backlog area of the Scrum Board. Where the Product Backlog is a listing of all the stories that might be done for the entire project, the Sprint Backlog is a list of stories to be done in the current sprint. Any story not completed, or any new story realized goes into the Product Backlog for consideration.

5.3 Identify Tasks

	Process Objective	Activity	Artifacts	Meeting
6	Identify Tasks	Identify the tasks that need to be done to meet Product Goals complete with measures, targets and tests that define when they are done (i.e. test-driven development).	Tasks for Sprint Backlog and Product Backlog	Sprint Planning Meeting

Where the stories describe what the customer wants, the tasks are what the team needs to do to complete the stories. It is best to write tasks on cards similar to those used for the stories, but with different fields, Fig. 13.

Task						Story ID	
ID		Owner		Time estimate		Actual time	
Description							
Measures and targets							

Figure 13 Task card

On the first line is the name of the task and the ID number for the story it supports. The task name should be something descriptive of the deliverable, the result of completing the task. Each task needs an ID of its own, and the owner is the person (or people) on the Technical Team who will be completing the task (see below for how they are determined).

In the middle of the card, there is a field for a description of the goals of the task. At the bottom, these are further refined as measures and targets. A task has clear deliverables[iii] i.e., something to show when the task is complete. These can range from an analysis that shows how well test results match a target, the confirmation that a drawing is complete, a list of literature read or any other tangible evidence that the task is completed.

Tasks should have the following format:

> The team will <activity> to achieve <measurable deliverable>

Examples:
- The team will develop sketches for four different mounting concepts.
- The team will analyze the OBOGS documentation to determine what inputs it needs. The team will develop code to compute the mass of the OBOG.
- The team will build a prototype of the gripper sufficient to demonstrate that it can pick up a 12oz. Dixie cup full of hot coffee after it has been filled for 5 minutes.

Some typical activities are listed in the sidebar.

It is important here to think in terms of Test-Driven Development (TDD). The task's measures and the targets for them define the tests needed to confirm the task is complete. For mechanical design, this forces the granularity of the tasks to be quite small. While the Scrum methodology does not offer any structured methods for developing the "tests," the QFD[iv] method does map the voice

Task activities:
- Develop
- Analyze
- Choose/select/decide
- Explore
- Integrate
- Build
- Test
- Validate
- Investigate
- Demonstrate
- Write

of the customer to measurable engineering specifications. These specifications define the "tests" needed for TDD. The important takeaway here is that, for good practice, design the tests as you design the product. This may include the design of the experiments, the test equipment, and the data reduction system.

Fig. 8 showed an example of the requirements for an aisle chair using the QFD method. One requirement, in story format, was: "As an airline agent I would like to be able to adjust the seat height easily to make a safe transfer of the customer from their personal chair to the aisle chair." One measure of this found with the QFD was "Steps to adjust seat height" with a target of 3 steps. In other words, one way to tell if the story of adjusting the seat is complete is to measure the number of steps needed to adjust it and that they are less than 3. This task might look as shown in Fig. 14.

Task	Develop concepts for seat height adjustment and rate them			Story ID	1.1.1
ID	4	Owner	Bob Smith	Time estimate	12	Actual time	

Description
Explore concepts to adjust the seat height

Measures and targets
• Detailed sketches of three methods to adjust seat height with evidence that: 　• Less than three steps to adjust 　• At least four different heights between 40 and 50 cm 　• Force needed to adjust without person in seat less than 50N 　• Less than 10 parts for adjustment mechanism

Figure 14 Example task

The student group at Olin used a briefer form of task description on their web-based project management system as taken from their Trello log (Fig. 15). Here each task is a simple statement of what needs to be done. They also have been able to add a link and an image to the topmost task to show progress on the task. What is lacking here is any sense of what "done" means. Just because a prototype has been built, does not mean that it works, does not tell how well it works, and does not allow the next version to be driven by any test results.

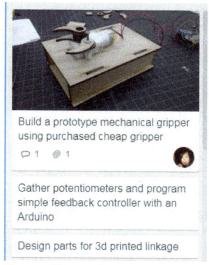

Figure 15 Example tasks from the Olin prosthetic arm.

5.4 Time Estimate Each Task

For each task, an estimate the time (person hours) needed to complete it is entered on the card (Fig. 14). Ideally, tasks should be 2 to 16 hours in length. Anything that will take longer should be divided into multiple tasks. Task estimate accuracy improves as the team gains maturity. Some methods for estimating time are in Section 5.4.3.3 of *The Mechanical Design Process* and a simplified version of one of these methods presented here.

	Process Objective	Activity	Artifacts	Meeting
7	Estimate Task Time	Estimate the time each task will take.	Tasks for Sprint Backlog and Product Backlog	Sprint Planning Meeting

Time estimation is notoriously inaccurate. One approach is to make an optimistic estimate (i.e., everything goes as good as it can) a most likely estimate and a pessimistic estimate (i.e., everything goes wrong). Calling these three estimates "o" for optimistic, "m" for most likely and "p" for pessimistic, then the time estimate is:

$$T = (o + 4*m + p)/ 6$$

This method is commonly used to take into account uncertainty.

Enter the best possible time estimate on the task card (Fig. 14). When the task is completed, the actual time is also entered on the card. Comparing the time estimate with the actual time will prove to be helpful in estimating future tasks.

As a reality check add up the estimates for all the tasks needed to complete each story and compare them to the original story estimate. The summed task estimate is generally more accurate than the original story estimate. If close to the story estimate, then all is well. However, if the summed task estimate is much higher, then one of the following needs to happen:

Ensure all the tasks are needed to complete the story. If not, then the stories and tasks may need to be rethought. Don't hesitate to push some stories or tasks back to the Product Backlog and address them in a future sprint.

The story may need to be broken into sub-stories. One or some of these will be completed in the current sprint, and the others will fall back into the product backlog and considered for future sprints. The number of stories to be addressed during the sprint (as chosen in Step 5) may need to be revisited and revised downward. The team may need to push an un-started story back into the Product Backlog.

There may be too much uncertainty in the story. If so, break it into many tasks where some of the tasks focus on reducing the uncertainty. In other words, there may need to be tasks focused on the requirements, and then tasks focused on concepts, and finally, when enough is known, tasks that address the product itself. The Olin College team used many early tasks to learn enough to then begin to design product.

5.5 Choose Sprint Tasks

	Process Objective	Activity	Artifacts	Meeting
8	Choose Sprint Tasks	Choose which tasks to complete during Sprint. Only start what you can finish.	Tasks for Sprint Backlog and Product Backlog	Sprint Planning Meeting

Since the Scrum method his timeboxed it is important only to choose to start the tasks that can be completed in the time available. Thus, the story time estimate made in Step 5 needs to be revisited and the tasks to undertake during the sprint chosen. Do not hesitate to push some stories and tasks back to the Product Backlog. At this point, you should only have what you think you can finish in the Sprint Backlog.

If a story has five tasks and you can only do three of them during the current sprint then the tasks to be done will be in the Sprint Backlog, and the other tasks and to story itself will be in the Product Backlog. A story is only completed when all its tasks are done.

5.6 Align Team Members with Tasks

Process Objective	Activity	Artifacts	Meeting
9 Align Team Members with Tasks	Choose what tasks to complete during the Sprint. Only start what you can finish.	Tasks for Sprint Backlog and Product Backlog	Sprint Planning Meeting

The Technical Team self-organizes to best apply the members' technical skills to the tasks. Note that the title of this task includes the word "align" and not the word "assign." An agile principle is giving the team autonomy over how it solves problems. In the ideal team, there is more than one member who can do each task and the Technical Team determines the best allocation of each member's time. Anyone who tackles a task can get help by requesting it in the Daily Standup Meeting during the design cycle or at any other time since the team is collocated. At any point in the sprint, Technical Team members can rearrange themselves as needed (see Steps 10 and 11).

This sprint structure keeps the number of parallel tasks to a minimum reducing multitasking and the loss of efficiency that comes with the lack of focus.

In highly technical fields the tasks may be assigned to individual team members with sufficient expertise to resolve the issue. This is often the case at Saab as can be seen in the case study included later in this book. Also, the Olin prosthetic arm team was more typical of multi-disciplinary teams. Victoria worked on the software, and she and Celine identified and implemented the sensors. Liani and Ellie, both mechanical engineers, focused on the physical components and their manufacture. Still, the team members decided who was going to work on which tasks, keeping with the Agile Principles.

6. DO - THE DESIGN CYCLE

After the Sprint Planning Meeting, the technical work begins, and meetings kept to a minimum. "Doing" means that specifications are refined, concepts and products developed, and tests and analysis completed. Whatever is defined by the tasks. To manage this technical work, there is a Daily Standup Meeting which is very short and a Scrum Board to keep track of who is doing what as the sprint progresses.

6.1 Do Work

Process Objective	Activity	Artifacts	Meeting
10 Do Work	Do the technical work moving tasks from "To Do," to "Doing" to "Done" on Scrum Board.	Scrum Board, Product Artifacts	Sprint Standup

As a team member begins work on a task, they move it from the Sprint Backlog, the list of "To Dos" to "Doing" on the Scrum Board. As they complete a task, they share the results with the Product Owner. If the PO is satisfied that the results complete the task, it is "Done" and moved to that area of the Scrum Board. If not, then the team member needs to continue to work on it. Care must be taken that "done" is consistent with what was defined during the Sprint Planning. If more features are wanted, then a new story or task needs to be written and added to the "New Stories" area as shown in Fig. 9. Stories and tasks here are considered during the next Daily Sprint Meeting or next Sprint Planning Meeting.

The PO accepting the task as complete is the last step that brings the story full circle in Fig. 6. The team member now works with the team to decide the next task during the next Daily Sprint Standup meeting.

One characteristic of most agile methods is the Daily Sprint Standup meeting. It is just as its name implies. It happens every day, and everyone is standing up. This daily meeting is crucial to the method. During it, the team members report on their progress and regroup to ensure the work on the Sprint Backlog gets done.

Typically, these meetings are held to less than 15 minutes first thing in the morning. They are scheduled at the same time each day, and every member of the team must attend. During the meeting, each member of the Technical Team answers three questions:

- What did I do yesterday to help finish the sprint?
- What will I do today to help finish the sprint?
- What obstacles are blocking the team or me from achieving the sprint goals?

The goal of these questions is not to judge anyone's work, but to find what is and is not going well. If there are obstacles, then the Scrum Master is responsible for making them go away. If there are technical issues slowing down a task, then the team needs to swarm to help overcome the problems. The term "swarm" is used in direct analogy to a rugby game where the team groups around the ball carrier to help move the ball down the field. During a sprint, it is better to do fewer tasks and swarm to finish those that have been started. A half-done story is not done at all.

This meeting is "planning in the small." With traditional methods, there is a large planning effort at the beginning of the project and periodic large design review meetings. Here the team re-plans their sprint every day.

While Saab and most other professional groups hold daily standups, the students at Olin could not do so. However, beyond the twice a week class meeting, the Scrum framework encouraged them to meet more often, which they did at least four times a week. It also helped them structure the meetings to answer the three questions itemized above.

Managing evolving tasks is the core of the daily standup and gives the Scrum framework the ability to react to uncertainties. Design is learning, and learning means that new tasks are realized and must be managed or too much time can be spent pursuing the wrong things. This introspection by the team each day is another powerful feature of the Scrum framework.

6.2 Track Progress

	Process Objective	Activity	Artifacts	Meeting
11	Track Progress	Use Burndown chart to track progress.	Update Scrum Board	Sprint Standup

There are two ways to track progress during the sprint. First is the movement of tasks and the stories they support across the Scrum Board from "To Do" to "Done". This assumes that the team must easily see the Scrum Board, so they tell at a glance how the sprint is going. Most companies have the Scrum Board on a prominent wall. The Olin student team could easily see theirs on the web.

Another tool to map progress is the Burndown Chart (on the right in Fig. 9, the added bar chart taped on in Fig. 10, or the detail in Fig. 16). The team can easily track their progress on this chart. For example, in Fig. 16, during the Sprint Planning Meeting the team decided that the tasks chosen should take 270 hours to complete. For the 20-day (4week) sprint, if they made steady progress, their burndown would follow the straight line from 270 hours on day zero to zero hours on day 20.

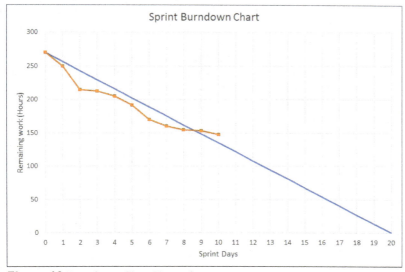

Figure 16 Burndown Chart Example

Each day the Scrum Master logs the team's progress by summing the hours of the done tasks and subtracting them from the total. As can be seen on the example plot, the first day the team actually accomplished twenty hours of the work, so there are 300 hours of work remaining. As the days go by, the team gets ahead early on and then makes slower progress on days 7-10, falling behind.

If this rate continues, they will have some incomplete tasks at the end of the sprint. According to Scrum orthodoxy, <u>a task that not completed during a sprint has no deliverable and is of no value to the customer</u>. So, the team must decide during one of its Daily Standup Meetings what to do about the slipping schedule. They have many options:

1. Work faster or longer hours - not a good idea. First, it violates the ten Agile Principles driving down motivation and reducing collaboration. Second, it is better to work smarter, not harder and agile makes this possible.
2. Add more people to the project. Adding people takes time for them to ramp up, increases communication channels and disturbs the team cohesiveness. These issues are so common that Brooks' Law (in the text box) has become well known in the software industry. While overstated, adding people is seldom a good response[5].

Adding a person late in a project, makes it take more, not less time.
Brooks' Law

---

3. During the standup meeting, the team can decide to rewrite a story or task that is problematic or even break a story into sub-stories. They can then keep some of the sub-stories in this sprint and put the others in the NEW STORY section of the Scrum Board for inclusion in a later sprint. The story or stories kept in this sprint must still have tasks written with clear deliverables. The philosophy here is that the team can rescope their work, so they are "done enough for now." This is an admission that the story or task was too large for the sprint and there is an identifiable middle point that may be sufficient for the project or is at least sufficient until the function is revisited in a later story. Understanding what is "done enough" is an essential part of engineering maturity. It comes with experience.

If the team realizes it is ahead, that the actual progress is better than expected and that they are going to get all the sprint backlog completed during the sprint, they can then go to the Project Backlog and pull another story or task into the sprint as long as it can be completed in the sprint time available.

At the end of the sprint, there are two meetings used to assess the sprint: The Sprint Review focuses on the product, <u>what was accomplished</u> during the sprint; and the Sprint Retrospective, focusing on <u>how the work was accomplished</u>.

---

5 The anthesis of Brooks' Law is the "Bermuda Plan" – send 90% of the engineers to Bermuda, and the remaining 10% will get the project done. Also, overstated.

7. REVIEW

A key feature of the Scrum process is the reviews of both the product and the process at the end of each sprint.

7.1 The Sprint Review

	Process Objective	Activity	Artifacts	Meeting
12	Review Sprint Product	Hold a sprint review (aka design review) where the progress on the product is demonstrated.	Updated Product Intent and Tasks	Sprint Review

The goal of the Sprint Review is to demonstrate the technical accomplishments achieved during the sprint. In the non-Scrum world, this is called a design review. The team and any others outside the team may attend the Sprint Review Meeting such as customers and marketing representatives, or members of other teams. This meeting is a go, no-go decision point. Each story or task has been completed, or it hasn't. If not completed, then a new story or task is written and put in the Product Backlog.

The Sprint Review Meeting is not a long meeting, timeboxed at one hour. If the voice of the customer has been well communicated through the Product Owner and well represented in the Daily Standups, there should be no surprises and no need for a long meeting.

7.2 The Sprint Retrospective

	Process Objective	Activity	Artifacts	Meeting
13	Review Design Process	Have a Sprint Retrospective where the design process is reviewed, and improvements developed.	Team Process Changes	Sprint Retrospective

In the Sprint Retrospect Meeting, the team focuses on the sprint process itself. To prepare for this meeting the Technical Team members write on sticky notes short descriptions of "Things that went well, that we should continue doing," and "Things that didn't go well, that we would like to be better in the future." They write one item per sticky note and stick the notes on a white board under two columns: "Things That Went Well" and "Things To Improve" as shown in Fig. 17.

The Scrum Master conducts the meeting. It is a search for how to improve, not a witch-hunt for the guilty. This meeting is a critical element to help teams mature.

Teams improve over time in their ability to estimate the time it takes to complete a story or do a task, in deciding who should work on what, and how to address issues that come during a sprint. By keeping a team together over a long period, which is very common, this maturity can improve the product quality and amount of design work possible. Some teams at Saab have been together for ten years.

At Saab some of the ideas for improvements go directly into the teams' backlog; other ideas communicated to other teams, for example, support teams; yet other issues escalated to management.

Figure 17 A Typical Retrospective

It is important to realize that Scrum is a "cookbook," which makes it easy for the team to implement activities without any reflection on how well they work. The retrospective allows evaluation of what parts of the team behavior are beneficial and suit the specific organization's needs.

Experience in industry has shown that teams improve through these retrospectives through increased speed in developing products, better ability to estimate the time needed to do tasks and overall team efficiency.

8. START NEXT SPRINT CYCLE

The review meetings bring the cycle back to the beginning (see Fig. 6) where, for the next sprint cycle, the team addresses the new, most critical stories with a Product Grooming Meeting and Sprint Planning.

The Olin student team completed four sprints in their 8-week course. The results of Sprints 1, 2 and 4 are shown in Fig. 18. The evolution of the product is clearly evident: from the crude gripper in sprint 1 to the final hand (better detailed in Fig. 1): from the wooden box frame to the fore-arm shape, and from no circuit in Sprint 1 to the packaged electronics in the final product.

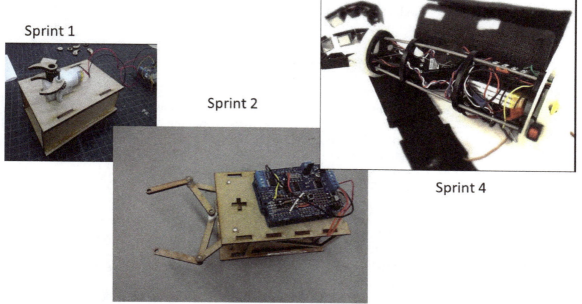

Figure 18 The results of the Olin sprints.

There was much uncertainty at the beginning and learning throughout the sprints. A complete case study of the team's progress through the design process is in the next chapter of this book. Also included is the case study of a team at Saab aerospace who has been together for ten years. They still deal with uncertainty in every project, and the Scrum methodology continues to help them develop quality products.

9. MIXING SCRUM AND WATERFALL

Where most hardware development companies rely on the traditional waterfall processes, leading-edge companies are increasingly moving to Agile/Scrum methods or a mix of both methods.

> Organizations are constrained to design products which are copies of their communication structure.
> *Conway's Law*

Waterfall works best when:
- Products are mature with little uncertainty – situations where new products are very similar to existing designs.
- Product specifications are well known, can be developed or must be developed due to the size or complexity of the system or the demands of the customer.
- They can be plan-driven with most of the scheduling and forecasting up front.
- There is a focus on product time to market and cost.
- The company is siloed (i.e., technical functions in separate organizations) with poor communication between marketing, design, and manufacturing. Siloing is known to be inefficient, yet many companies still support this structure.
- The product development team is distributed, maybe even in different countries, in different time zones.
- The organization demands top-down management.

Agile/Scrum works best when:
- Adaptability is needed.
- The product specifications are co-developing or being refined with the development of the product.
- There is no detailed upfront plan. Emphasis is on detailed short-term planning.
- The focus is on product function. Time is managed by the inclusion or exclusion of function.
- The team is collaborative representing the product from marketing to manufacturing and beyond.
- All team members can be collocated. There are some workarounds for this as described in the next section.
- The organization can support project management by teams.
- Saab Aerospace and other leading product design companies are mixing the two strategies.

Combining Agile and waterfall can result in:
- Faster response to change
- Improved team communication and coordination
- Improved productivity
- Higher morale in project teams
- Improved project focus

For the design of the Gripen E fighter, Saab spent three years planning the airplane in a typical waterfall method with Gantt charts, but the detailed design work is all in sprints.

The transition from pure waterfall to the inclusion of Scrum can be difficult and may not be well received. Change comes slowly to organizations as is well stated in Drucker's Cuckoo Effect. It is best to transition slowly while structuring the project at the highest macro levels as a waterfall, and the details as sprints.

> Any innovation in a corporation will stimulate the corporate immune system to create antibodies that destroy it.
> *Drucker's "Cuckoo Effect"*

Research has found[v]:

- Integrating Agile with waterfall processes **make the planning more efficient**. Inflexible and fixed plans are avoided so that the risk of project delays minimized.
- The **continuous integration of functionalities provides faster, and earlier feedback loops**. It can eliminate the final testing of the entire product at the end of the project.
- **Resistance can emerge among the managers** because of losing control over the Agile teams.
- **Tasks are split into smaller tasks** which facilitates the estimation of how much work can be done within one iteration and the time it takes to estimate the work was decreased.
- The teams **perceive improved internal team communication**, and they feel that they have control over their work, which increases the motivation of the designers' daily work.
- **Face-to-face communication is more effective than communication through documentation**. The project status is easier to visualize and clearer than in the traditional process.
- The combination of waterfall and Agile methods results in better response to changing customer needs, more proactive and effective ways to build in customer needs, improved ways of dealing with resource issues, reduced cycle time and higher productivity.

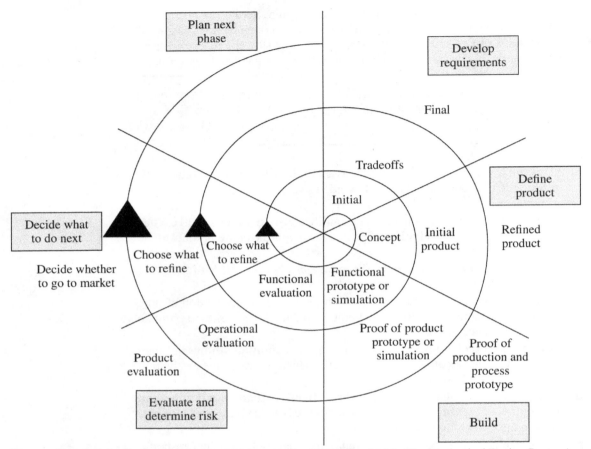

Figure 19 Spiral development of mechanical Systems (reprint of fig. 5.15 in The Mechanical Design Process)

Another way to look at sprints and their integration into waterfall is that they enable Spiral development. Fig. 19, a reprint of Fig. 5.15 in *The Mechanical Design Process*, shows the spiral process. Design begins in the center and as the product matures the process spirals out through all the phases noted. The spiral process was proposed to help offset the waterfall limitations. Its development predates Scrum and is part of the evolution toward it. This evolution is easy to see when you visualize each loop of the spiral as a sprint or multiple sprints reducing uncertainty and driving learning. The case study results in Fig. 18 show a spiral of learning.

SOURCES

i Bill Wake, "INVEST in Good Stories, and SMART Tasks," Posted on August 17, 2003, https://xp123.com/articles/invest-in-good-stories-and-smart-tasks/

ii Section 5.4.3 in *The Mechanical Design Process*, 6th edition addresses sequencing and other task estimating methods.

iii Section 5.4.3 in *The Mechanical Design Process*, 6th edition includes details on defining tasks.

iv Chapter 6 in *The Mechanical Design Process*, 6th edition, addresses Quality Function Deployment.

v Karlström and Runesson (2005) "Combining Agile Methods with Stage-Gate Project Management". IEEE Software 22(3): 43–49.

APPENDIX A: A GLOSSARY OF AGILE/SCRUM TERMS

- **Agile** - the ability to respond effectively during the design process to uncertain, risky, variable, and evolving information.
- **Agile Principles** – Ten basic principles that guide all hardware agile methods.
- **Burndown Chart** – a tool used to plot progress during a sprint – see Fig. 16.
- **Customer Story** – see Story
- **Daily Standups** – See Sprint Standup
- **Design Cycle** – The "Do" portion of a sprint where design progress is made – see Fig. 4.
- **Goal** - The product needs in terms of the voice of the customer and captured through stories and requirements.
- **Hardware** – Any artifact being designed that is not software including mechanical, electronic and mechatronic systems.
- **Product Backlog** – an area on the Scrum Board where stories and tasks are parked waiting to be addressed in a sprint – see Fig. 9.
- **Product Backlog Grooming** – a meeting where the product backlog is refined- see Section 5.1
- **Product Owner (PO)** – The voice of the customer on the team – see Section 4.1.
- **Requirement** – a design goal used here to imply it comes from the use of QFD or other non-story.
- **Scrum** - an agile method initially developed for software design – see Section 1.
- **Scrum Board** – a tool used to manage a sprint that can be physical, on the wall, or virtual, online – see Section 4.3.

- **Scrum Master (SM)** – a team member who drives the process and is accountable for removing impediments, so the team can efficiently complete the design cycle – see Section 4.1
- **Spiral Process** – an iterative design process well supported by a series of sprints – see Section 9.
- **Sprint** – A Timeboxed Design Cycle – see Section 2.
- **Sprint Burndown Chart** – A tool for tracking progress during a sprint – see Section 6.2 and Fig. 16.
- **Sprint Backlog** – A list of the tasks and stories to be addressed during the current sprint – see Section 5.2.
- **Sprint Planning Meeting** – Plan the tasks and stories to be resolved during the sprint – see Section 5.3.
- **Sprint Retrospective** – A meeting at the end of a sprint to review and improve the process for future sprints – see Section 7.2.
- **Sprint Review** – A design review at the end of a sprint – see Section 7.1.
- **Sprint Standup** – Daily sprint meeting to review what was done yesterday, what will be done today and identify problems – see Section 6.1.
- **Standup** - See Sprint Standup.
- **Story** – A method to represent design goals used in software engineering and helpful here – see Section 4.2
- **Swarming** – The ability of a team to reallocate resources during a sprint to resolve difficult issues – see Section 7.1.
- **Task** – The activities that need to be done during a sprint to complete stories – see Section 5.3.
- **Team** – The Technical Team, Product Owner and Scrum Master who execute a Sprint – see Section 4.1.
- **Technical Team** – The people who do the technical work during a Sprint – see Section 4.1.
- **Test Driven Development (TDD)** – The concept that every task should have measurable targets that can be tested to prove they have been completed – see Section 5.3.
- **Timeboxed** – A predefined length of time for a Sprint or a meeting – see Section 2.
- **Waterfall** – A sequential design process common in industry – see Section 2.

APPENDIX B: HARDWARE CHALLENGES

Software design has widely adopted Scrum methods. Their application to physical systems has been much slower. There are quite a few reasons for this and, while they all cannot be overcome, many Scrum concepts make physical product design more efficient and can lead to higher quality products in the end. This section lists the factors that create challenges and opportunities for Agile/Scrum application to the hardware design process.

1. Modularity is the key to agile.

Software teams have long understood modular architectures[i] and have been using them to limit the ripple effect of each change. In software, modular architectures allow an application to be broken into chunks of code, each of which provides a small set of functions. Each of the modules can be developed as a virtually independent design project. Where form and function are often interdependent in hardware, in software there is no actual physical form. The code, the closest thing to form, is the function. For the most part, this allows software to decompose functionally into narrow vertical slices that can be developed in small updates, nearly independently.

In hardware systems, the form and function boundaries[ii] overlap leading to a more challenging ability to modularize. There can be a ripple effect when one module is changed, and other modules share in producing the function. These effects can be controlled, to a large part, by designing around known, stable interfaces between modules. As the design matures from constraints to configurations, to connections and finally to components[iii], interfaces can be designed that break the form-function interrelationship. This design philosophy forces the configuration or architecture of the product to be established early on defining functional modules. Then the connections or interfaces between the modules need to be designed first. These become the known stable interfaces between modules.

For example, Saab during the design of their new Gripen E fighter aircraft has managed to implement agile practices in hardware design. They overcome the difficulties mentioned above by focusing on the interrelationships early in the process. The aircraft is divided into teams aligned with the project organization. As the design progresses from concept to detailed design, teams work almost autonomously within known stable interfaces.

The trick is to define interfaces and make them as simple and stable (i.e., unchanging) as possible. This way the design effort can proceed independently on each side of the interface. Later, the interface can be redesigned if needed as a separate project or as part of the ongoing Scrum activity. Making modules plug-and-play with these known stable interfaces is the goal encouraged by the Scrum framework.

2. The Scrum methodology operates around short design sprints of 2 to 4 weeks.

The ideal goal of each sprint is "a potentially shippable increment of product functionality." This ideal is difficult to achieve in software and even more challenging in hardware. In fact, many software groups set more modest, yet demonstratable acceptance criteria that are not actually shippable. The setting of realistic sprint task acceptance criteria is one key to the use of Scrum for hardware design.

Where software development builds on the aggregation of small functional deliverables over time, hardware often cannot. Hardware more commonly builds up a set of components and assemblies that address a target level of functionality. Thus, the goal of a hardware sprint may not be "functional" but may focus on the design of components or assemblies.

Sprint success is all dependent on setting realistic sprint goals. Where the ideal is a shippable increment, the reality may be the team demonstrating that: "we understand the question," "we have a demo of this function" or "we have a working prototype of the feature." The challenge is in writing sprint stories with acceptance criteria that describe "we are done enough for now." In other words, the goals should describe small bites that lead to "a potentially shippable increment of product function" and that are achievable during the sprint cycle.

3. Software products evolve through multiple releases by adding new functional features.

This is seldom possible in hardware design. In fact, "feature creep" (adding new features during the design process[iv]) can be fatal during hardware development. In software, it is relatively easy to add and subtract features to meet evolving requirements. However, in hardware, there are often unintended consequences due to the form/function interaction.

To counter feature creep, set both the software and hardware high-level requirements in advance. The Saab Gripen fighter was well defined years before it was completed, and these requirements filtered down to sprint stories. The requirements are defined at an early stage but continuously reviewed.

4. Software is often refactored, rewritten to be made simpler.

If a particular algorithm chosen for a software function is inefficient, it is not expensive to rewrite the code. Only the coding time is lost. Refactoring physical matter is usually much more difficult and expensive than refactoring software.

In a sense refactoring is the software road to product maturity. Hardware matures from concept to product, and then from the first version of the product to later versions that are more efficient, weight less, are less expensive to produce, and so on. Maturity evolution is enhanced with the careful definition of the product's modules and their interfaces. The more a product has plug-and-play assemblies and components, the easier it is to mature and the more it can be refactored.

5. Hardware design requires more specialization than software design.

While there are multiple languages and areas of expertise in software, this is nowhere to the degree that it is for hardware. For software, a person who can code function A probably can code function B. In hardware systems, expertise is not universal across functions. A person who can design a heat exchanger probably has no experience in the design of gear trains or printed circuit boards. While the effort to utilize people with "T" expertise (great depth in one area, but breath into others) is ideal, it is not always possible. While software assignments are often made during the Daily Standup Meeting, at Saab and other companies, a specific engineer is often assigned during sprint planning.

While hardware and software often need specialization, specialization causes bottlenecks. If there is only one person who can do X and three teams need X, there may not be enough time for each, and the Scrum concept of stable teams comes apart. Additionally, specialization is fragile because what if the expert quits or becomes ill, the capability is lost. To counter specialization, many companies who cannot afford to hire many people with deep expertise try to spread the knowledge by having the experts train others to do much of the lighter work and only use the experts for very deep problems when necessary.

6. The time needed to demonstrate function is higher for hardware than for software.

Code can be written and compiled very rapidly; designing and building physical objects take more time. Additive manufacturing (aka rapid prototyping or 3-D printing) and computer modeling are reducing the time to produce hardware that can demonstrate functionality.

Some hardware systems can only be simulated before final assembly, think of a space station or aircraft. Improved simulation capabilities are enabling digital twins, simulations that function identically to the final, assembled product. These enable changes to be studied without the need for hardware and thus much more rapidly.

7. The cost of change gets higher for hardware as the project progresses.

For physical systems, as the design process evolves, cost is committed that cannot be recovered[v]. Not only is the cost of salaries lost, but tooling, inventory commitments, and other sunk costs are lost for changes made after a certain point. The cost of changing software is only the cost of programmer time regardless of when in the development process. This difference is hard to overcome and is a reality of life for hardware. The iterative nature of sprints with goals of "good enough for now" forces considering the potential for a need for change downstream and the effort keep it under control.

8. Hardware must work over a range of time and environmental conditions which is not the case for software.

Software does not bend or fatigue, nor does it care if it is cold or hot. However, software cares what machine it runs on, what operating system is used and often has difficult legacy issues (e.g., version 2 must process data from version 1 while adding new functionality). Thus, software and hardware both have "environmental" issues just not the same ones.

9. Software testing is very different than hardware testing.

Testing is to reduce risk by proactively finding and eliminating problems that can affect the product's usability. Good story acceptance criteria define tests that need to be passed to complete the story. Software testing usually includes thousands of test cases, so programmers write specialized code to root out potential problems. Thus, software is tested by designing more software.

Hardware usually requires fewer test cases but with more specialized and expensive equipment. This testing equipment itself may be designed and built as part of the project.

Regardless of whether designing software or hardware the Scrum focus on test-driven development is very important to the design of quality products.

10. Hardware development requires the design of 1) the product, 2) the manufacturing process, 3) the test equipment, 4) the supply chain, 5) and the documents and their management.

For software, writing code is the manufacturing process that generates the product, there is no supply chain, and the documentation is generally simpler than for hardware systems.

One approach to managing the complexity, in either case, is to bring all these "design" issues into the team and have people who can design the manufacturing process, the supply chain and the other needs on it.

For physical products, it is well known that when manufacturing has a role in product design, the design is easier and less expensive to manufacture. It is also well known that you can't inspect in quality into a product, so having the team design quality it in is critically important. Finally, it is far easier to document a product as it is being designed, than after design is finished. All these factors encourage having people on the design team who can address tasks focused not only on the product but also on manufacturing, testing, quality, procurement, and documentation.

11. User stories are a big part of software development.

Resolving user stories drives sprints. User stories are outward - customer facing. Hardware is more inward - technology focused. Technical stories are the same as user stories rewritten in the form: "As a <system> I want to <perform an action> so that I can <gain this benefit>." Many argue that technical stories only exist to provide function to a user. In any case, methods like Quality Function Deployment[vi] provide a method to generate a rich set of user or system stories.

12. Demonstrating, prototyping and testing are often more difficult for hardware than software.

Working code is easily tested by customers, even globally over the internet. Hardware is not. It is undoubtedly true that many hardware issues can be addressed quickly with the CAD models and that rapid prototyping and printing (additive manufacturing) have significantly reduced the time it takes to produce a physical prototype. However, when these technologies are not sufficient, there may be lead times on materials, manufacturing and test facilities.

13. Agile encourages build-measure-learn.

This action sequence is what we naturally do when faced with a problem. Watch any kid learn a new skill. They will try something, watch it fail and, through the failure learn enough for the next cycle. The adage "fail early, fail often" has real truth to it. However, take care as agile methods can lead to poor decisions and a premature commitment to a weak concept if not very carefully done. History is littered with instances where early builds with poor measurement have led to erroneous learning and a lot of wasted time.

While build-measure-learn is a good process, take care in the development of requirements and concepts. When designers build without fully understanding how they are going to measure and learn, they are hacking or writing "cowboy code," not doing engineering design. When designers commit too soon to a concept and start to build it, they are often wasting time and resources.

With short sprints, daily standups and sprint reviews every two to four weeks designers get rapid feedback. This near-continuous feedback helps curb premature commitment and hacking. Designers get to prove or disprove hypotheses as fast as possible.

Sources

i Section 7.9 in *The Mechanical Design Process*, 6th edition.
ii Section 2.4 in *The Mechanical Design Process*, 6th edition.
iii Section 9.3 in *The Mechanical Design Process*, 6th edition.
iv Section 6.1 in *The Mechanical Design Process*, 6th edition.
v Section 2.2.1 in *The Mechanical Design Process*, 6th edition.
vi Chapter 6 in *The Mechanical Design Process*, 6th edition.

Agile Design of an Agile Fighter at Saab Aerospace

A CASE STUDY FOR THE MECHANICAL DESIGN PROCESS

INTRODUCTION

The Saab JAS 39 Gripen is a light single-engine fighter aircraft designed using Agile methods. The E model, first test flown in 2017, can reach Mach 2.0 with a combat range of 1500 km (932miles).

Besides the Gripen's impressive flight characteristics, its development costs were about one-tenth that of the Lockheed Martin F-35 to which it is often compared. Saab achieved these cost savings while improving quality using the Scrum design process - an Agile methodology. This case study explores the use of Scrum for the design of the Gripen E's oxygen system and its testing during 2018.

Figure 1. The JAS 39 Gripen

The team featured here consists of 10 members and is known as the "Escape&Oxygen" team. It is responsible for pilot survival and integrating an oxygen concentrator supplied by Honeywell with an ejection seat supplied by Martin-Baker.

- **The Problem**: Develop the oxygen/escape system for the Saab 39-E through the integration and testing of components supplied by other vendors.
- **The Method**: Use Scrum to manage product evolution, testing, and team communication.
- **Advantages**: Saab's focus on autonomous teams using a Scrum design process both reduces bureaucracy and encourages decision making at the lowest possible level in the project organization. There is clear communication between individuals and teams leading to an aircraft delivered for lower cost, faster, and of higher quality.

BACKGROUND

The Gripen E is the latest Saab fighter. It was designed to be lighter and less expensive than many of its competitors such as the US Lockheed Martin F-35[i], the Russian MiG-35[ii] or the Chinese Chengdu J-10[iii].

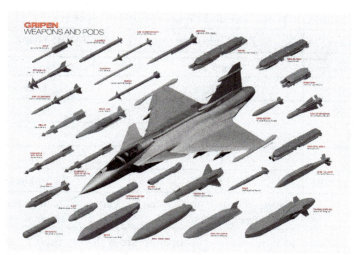

Figure 2. The Gripen with its optional ordinance

The Gripen fighter was first introduced in 1997. The E model is the latest in the series where A was the first single-seat version, B was similar to the A with twin seats, C was the second-generation single seat and D the second-generation twin-seat version. The E version has a single seat, and the F, still in development, will be a twin-seat version. The E is basically a new airframe, larger than the C model and with many changes to the oxygen system. As can be seen in Fig.2, the Gripen can carry a wide variety of weapons, and communicating and sensing pods.

A comparison of the Gripen E to the Lockheed-Martin F-35 Lightning, the newest US fighter, is shown in Table 1. Both aircraft have similar operating speed and range and can carry similar weapons. However, while the F35 is a stealth fighter, probably the best ever designed, the Gripen has a very low radar signal compared to earlier aircraft. Off-setting this stealth capability is that the Gripen is much less expensive to buy and to fly. Further, it was designed to be much easier to maintain.

Table 1. Comparison between the F35 Lightning and the Gripen E

Measure	F35 Lightning	JAS 35 Gripen E
Max Velocity	Mach 1.6	Mach 2
Combat Range	1300km	1500km
Stealth	Yes	No
Sensors	State-of-the-art	Good
Initial Cost	$94M -122M	53M
Operating cost	$31,000/hr	$4,700/hr
Maintainability	Needs a hangar and special equipment	Very good, can be repaired in the field

The low cost to purchase and to operate are a direct result of the design process used at Saab. As discussed in Section 2.2 of *The Mechanical Design Process*, 6th edition (*MDP6*) design can have a substantial effect on cost, quality and time to market and that is reflected in the Gripen E.

THE USE OF SCRUM AT SAAB

Like any modern fighter, the Gripen is highly complex. To manage this complexity, Saab introduced Agile practices such as Scrum, Lean, and Kanban in the early 2000s, during development of the earlier Gripen versions. The use of Scrum began with software design and spread throughout all the other aircraft disciplines. Since Agile methods encourage introspection and change, the methods applied at Saab have evolved over the years. By the time of the design of the Gripen E, Scrum had been implemented in every discipline and at every level in the organization.

The Gripen E was designed by more than 1000 engineers grouped into more than 100 Scrum teams[iv]. All teams have the same three-week sprint cycle, starting and ending at the same time. This gives Saab a common rhythm and a stable pulse. Saab has realized that Scrum methods provide:

- Clarity: Each team knows what is expected of them.
- Creativity: The teams know what is desired, not required as that would stifle creativity.
- Good decisions: Saab has driven decision making to the lowest possible level enfranchising teams.
- Clear communication of needs: While the ideal Scrum team interacts with the customer, this is not possible in a large project like the Gripen. To counter this, the Product Owners (POs) establish the value of features, set priorities and coordinate with other teams.

The project master plan for the Gripen E began in 2013 with the first customer contracts. As is common in Saab, this plan was broken into development steps with well-defined functional targets for each specific prototype aircraft. Each one has a test program with minimal functions for the first months, and additional functions added as the tests progress.

Development steps at Saab are broken down into "Increments." An increment is timeboxed at one calendar quarter, and this gives a well-defined delivery schedule. Increments are further divided into sprints with four three-week sprints in an increment. Functional increment targets are established each quarter and drive sprint planning. While orthodox Scrum requires delivery of product at the end of each sprint, Saab tries to have well- defined sprint targets, but these are not always deliverable.

This top-down organization of master plan, development steps, increments, and sprints are necessary on a project as complex as the Gripen. To manage this complexity, there are structured meetings to identify the system dependencies and make them visible across the project. These meetings are revisited at the beginning of each three-week sprint. System integration occurs not only at the end of each sprint but during them also as needed, ensuring that issues are visible and corrective action can be taken as soon as possible.

FOCUS OF THE CASE STUDY

This case study focuses on the design of the Gripen E oxygen delivery system. The E model has a single pilot who must be supplied oxygen throughout the flight envelop, up to a ceiling of 16,000m (52,000 ft). Further, the system must also supply oxygen in the event that the pilot ejects from the airplane.

Historically, early oxygen systems used Gaseous OXygen (GOX) with oxygen stored in metal cylinders at 1800psi (12.4 MPa). The weight and size of GOX systems increase proportionally with desired flight duration limiting their use in long-range flights.

More advanced systems used Liquid OXygen (LOX). These systems allowed greater storage since the oxygen was in liquid form which expands 900 to 1 in use. The liquid oxygen is not only pressurized but must be kept cold (-197F or -127C) and then warmed for delivery complicating the system and adding weight. Even with high expansion ratio, there is still a fixed amount of oxygen that can be carried, limiting the range.

Recently GOX and LOX have been replaced by On-Board Oxygen Generation Systems (OBOGS)[v]. With an OBOGS, the aircraft generates its own oxygen. Honeywell, at its British facility, makes the one used in the Gripen.

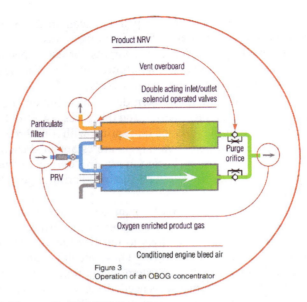

Figure 3. OBOGS' Operation

To explain how the oxygen is "made" Fig. 3 shows a schematic of the OBOGS. It uses what is called a "swing system" with two Zeolite beds (the center rectangles). Zeolite is a molecular sieve that lets oxygen molecules through and captures nitrogen molecules. There are two of these beds, one in use (the lower one in the figure) while the other is back-flushed to remove the nitrogen. When the lower one becomes nitrogen saturated the process is "swung" so the top one now is active and the bottom one is flushed. Conditioned (i.e., cooled) engine bleed air is brought into the beds (at the bottom of the figure) and oxygen enriched air (up to 95% oxygen) is produced (on the right). The remaining 5% is argon which does not affect human respiration. The oxygen-rich product of the OBOGS is later diluted down to a breathable level by the regulator/controller.

The OBOGS is integrated into the system as shown in Fig. 4. Oxygen flows from the OBOGS to a regulator/controller and then to the pilot. There is also a GOX Backup Oxygen Supply (BOS) integral with the ejection seat. In the case of ejection, the OBOGS in the airframe is disconnected, and the BOS takes over. The regulator/controller manages the source, pressure, and flow, and compensates for altitude.

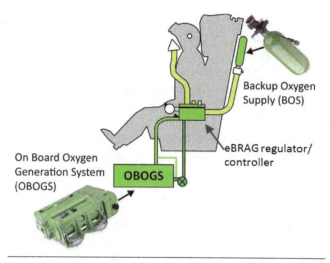

Figure 4. The OBOGS in the Gripen system

The regulator/controller, called the eBRAG manages the source of oxygen and the oxygen concentration. "eBRAG" means Breathing Regulator and Anti-G valve. The logic in this system combines pilot control with automatic sensing of flight conditions to provide a breathable oxygen mix.

Part of the challenge for the Gripen E Escape&Oxygen team is the need to integrate the OBOGS with the ejection seat. The seat is a Martin-Baker Mark S10L zero-zero (zero altitude, zero speed) rocket-propelled ejection seat[vi], Fig. 5. This is a well-tested seat with over 5,500 in service and over 800 lives saved.

The seat begins the ejection sequence when the pilot pulls the firing handle located between their legs. This initiates firing of a detonating cord which blows off the canopy. Then, the main gun located at the rear of the seat fires. This consists of a telescopic tube with two pairs of explosive charges that fire in sequence thrusting the seat clear of the airplane. As the seat moves up its guide rails, the OBOGS is disconnected, and the BOS is activated. At the same time, communication leads are automatically disconnected, and leg and arm restraints lock the pilot's limbs to the seat preventing injury.

As the seat moves further up and out of the aircraft, a rocket pack is fired. These rockets push the pilot and seat further from the aircraft. A small drogue parachute is deployed to stabilize the seat's descent path. A system prevents the main chute from opening above 16,000 ft (5,000 m). Once below this altitude, a time delay mechanism opens the main parachute. The seat then separates from the pilot for a normal parachute descent. A manual separation handle is provided should the automatic system fail.

With all this complexity, the Escape&Oxygen team is faced with a myriad of design tasks:
- Specify requirements
- Find suppliers and get quotations
- Evaluate standard products and specify changes
- Develop logic (data, alarms, etc.) Work with suppliers on the design Find optimal solutions
- Review/approve suppliers' test program Run Saab tests and reduce test data
- Develop mechanical and electrical connections between the supplied components and the aircraft

TEAMS

The one hundred plus Scrum teams at Saab who work on the Gripen E range in size from 5 to 10 members. Most of these teams have been together for a long time, some more than ten years. While each team generally has a consistent set of team members, they lose members due to normal transfers and retirements, and they gain members as new employees are hired. If a team gets too large, it may be divided into two teams, at the discretion of the team members themselves.

Figure 5. The Martin-Baker ejection seat

Within Saab, team practices are not prescriptive. Individual teams have the autonomy to develop the best implementation for their situation. Decisions are driven down to the team level whenever possible. Saab has learned that commitment and clarity drive performance and efficiency.

The Escape&Oxygen team has ten members: Peter is an expert on the eBRAG logic, Ingrid on testing, Gustav on the seat, Mattias on the software integration to the aircraft computer system, David on system safety, and others.

Figure 6. Escape&Oxygen Team and their Scrum Board

Like most teams at Saab, they all sit and work together to facilitate communication. Fig. 6 shows the Escape&Oxygen team and its Scrum Board. This photograph was taken at one of their daily standup meetings.

The Gripen E program uses the Product Owner (PO) as a proxy for the customer. The PO is responsible for establishing the value of features and works with all stakeholders on different management levels re- prioritizing them on the increment and sprint cadence. Each PO covers 1-8 development teams contrary to Scrum orthodoxy of one PO for each team. This is possible because there are relatively few Gripen customers and their voices are reflected through the high-level fighter aircraft requirements (See Requirements and Stories below). Also, this is necessary because the systems are so complex that one of the PO's primary role is coordination between teams due to systems' dependencies.

Each team has a team leader who acts as a Scrum Master holding the daily stand-ups helping the team decide who is doing what, resolving any disturbances to the sprint, ensuring all the needed input is at hand and so on.

A primary goal of the Escape&Oxygen team is the integration of components supplied by vendors such as Honeywell and Martin-Baker. These companies have their engineers in Great Britain. It is not economically realistic for them to have an engineer co-located in Sweden to be on the team. So, while representatives of these companies are not full-time members of Saab's Escape&Oxygen

team, there is a weekly Skype meeting with each vendor to resolve issues, and communication more often by email and phone is standard.

Since the ejection seat and oxygen system need to work with a pilot, Saab also has a pool of test pilots who are available for team integration on an as-needed basis. Many of these test pilots have engineering backgrounds and can integrate easily within the team.

REQUIREMENTS AND STORIES IN THE PRODUCT BACKLOG

For a fighter aircraft, the requirements develop over several years. They begin with the specification of the earlier version of the Gripen (the C model) and the changes needed for the desired operation of the E model. The requirements drive the product backlog for each team. The backlog contains the "stories" in bigger chunks, and sprint planning breaks them down to smaller pieces as tasks. While there are no "Stories" in the traditional Scrum sense, the goals that drive and focus team activities take many primary forms:

1. Requirements Development: Understanding airplane requirements and breaking them down to how they affect the new system.
2. Vendor selection: Solicit proposals, review proposals and choose suppliers.
3. Vendor specifications: Some stories require developing specifications for the vendors. These include determining design limits and other details that drive Honeywell's and Martin-Baker's design teams.
4. Detail design: Detail design in-house parts and assemblies and work together with suppliers developing their details.
5. Test specifications: The team often analyzes the airplane level requirements (derived from customer contracts and flight worthiness standards) to understand how they affect the oxygen system. For example, the aircraft level requirement requires that it can operate at -40 C. The team then analyzes test reports from the supplier and decides how to conduct their own "top-level" testing. This includes setting targets for passing or failing the vendor's equipment.
6. Test evaluations: The results of the tests are compared to the specifications and working with the vendors to overcome any issues,
7. Conceptual design: Make a conceptual design of hardware and software for both internal Saab systems and those supplied by vendors.
8. Quality control: Evaluate parts and assemblies and ensure they meet the specifications.

While formal Scrum practice suggests writing stories in the form "As a < (customer role or system)> I want to <perform an action> so that I can <gain this benefit>," Saab uses just simple headings. An unedited sample of Escape&Oxygen team stories from early in an increment are:

1. Define the mechanical interfaces between the Martin-Baker seat and the airplane.
2. Define the electrical interfaces between the seat and the airplane.
3. Define the testing program for the seat due to changes in the Gripen E (compared to Gripen C).
4. Define technical specification/performance of the Honeywell eBRAG/OBOG system with regard to oxygen concentration (up to 90% O2 at high altitude).
5. Define test program for O2 generation system.

The Product Owner prioritizes the product backlog with an interface to upper management and coordination with other teams' POs. This includes making sure the time plan for the upcoming sprints is correlated. For example; planning when the software integration is taking place, or what tests need to be performed together with other teams. Also, the PO must make sure that the product interfaces that are owned by other teams are coordinated. For example, the engine bleed air that provides the raw flow for the OBOGS (see fig. 3) is provided by another team.

TASKS

Tasks define the work done during a sprint. Ideally, a "task" includes the activity that needs to be done complete with measures for it and targets so that "done" is fully defined. Saab is very good at defining the activities to be accomplished, and during the sprint planning meeting, they discuss what needs to be accomplished. Then, during the daily standup meetings, they further discuss if the task is completed or if more work needs to be done.

The entire oxygen system is modeled in Simulink for testing functionality and as a solid model in CAD. Saab has developed models of most of the systems in their aircraft[vii].

The teams manage the tasks on Scrum boards as in Fig. 6. In orthodox Scrum, both stories and tasks are posted, but at Saab only tasks. The colors of the sticky notes in Fig. 5 indicate who on the team is responsible for the task.

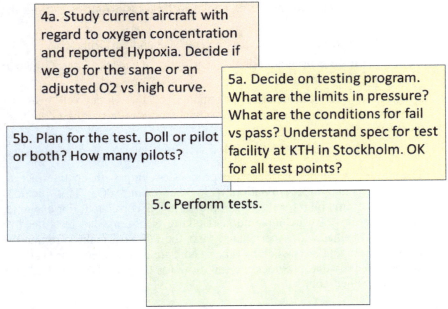

Figure 7: May 2018 Escape&Oxygen tasks on the Scrum board.

For example, the tasks addressed during the May 2018 sprint are shown in Fig. 7. These are from the "Doing" section of the Escape&Oxygen team's Scrum Board. Here the first number refers to the equivalent story in the story list, and the letters are just sequential numbering. Note that many of the targets are inferred.

CONNECTED BY STABLE FIXED INTERFACES

Modularity of design allows modularity of organization. In the design of hardware systems, modularity allows teams to work on their tasks relatively independent of other teams.

Fig. 8 shows the OBOGS with its major interfaces indicated by arrows. The yellow arrows point to pilot controls for the oxygen concentration and flow. These must be standard from aircraft to aircraft to ensure that a pilot has a stable environment in which to work.

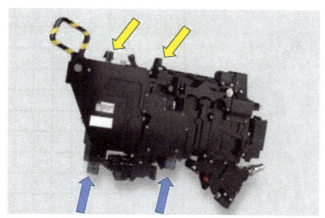

Figure 8. The OBOGS with interfaces indicated.

The blue arrows are OBOGS interfaces with other systems on the aircraft. These are where the oxygen-rich flow from the eBRAG is brought into the OBOGS along with other, low-oxygen air to dilute it. These are where signals from other aircraft systems are input so the OBOGS' logic can help keep the right mixture for the pilot. These interfaces are made standard early in the development of the aircraft along with the physical connecting points for mounting the OBOGS to the aircraft.

RETROSPECTIVE

Retrospective meetings are held at the end of each sprint. Both the Sprint Review and Sprint Retrospective are considered very important at Saab. The Sprint Review focuses on the product; what stories and tasks were completed, what needs work in a future sprint and when, and what new stories and tasks are needed.

The Sprint Retrospectives focus on the team process (how rather than what). Saab not only has team retrospectives, what could be improved in how the teams did their work, but also a retrospective of retrospectives across teams. This covers not only common problems across teams but also leadership and management issues.

Saab puts special emphasis on the Sprint Retrospectives with frank discussions of how the team worked during the sprint and generating more efficient ways to work in the future. During 2018 retrospectives, some of the ideas that the team developed were:

- A new CAD function was needed. It was defined and submitted it to a support team to implement.
- A weekly meeting coordinating with the test people was lacking so it was instituted.

- A larger Scrum Board was needed as the current board covered only one sprint and the team want a larger board to add in the rough contents of the two following sprints.
- An issue board (in Excel) was desired to support the meetings with Honeywell engineers complete with "status" updates.
- They found they needed a weekly software coordination meeting with stakeholders that are not part of the daily stand up.

All of these ideas developed during the Sprint Retrospectives were implemented.

CONCLUSION

Saab's focus on autonomous teams both reduces bureaucracy and encourages decision making at the lowest possible level in the project organization. On the product level, the result is an aircraft delivered for lower cost, with more advanced functions, and higher quality.

On a personal level: each team and each team member knows what is expected of them; they know what is desired, not required as that would stifle creativity; they make decisions and feel enfranchised, and there is clear communication between individuals and teams leading to product quality.

ACKNOWLEDGMENTS

Development of this case study was undertaken with the support of Jörgen Furuhjelm, Project Manager at Saab Aeronautics.

SOURCES

i   "Forget about the F-35, countries should be buying Saab's Gripen fighter jet", Air Force Technology, 15 Dec 2017,

ii https://www.airforce-technology.com/comment/forget-f-35-countries-buying-saabs-gripen-fighter-jet/

iii "How does the Saab Gripen E compare to the Su 35, Su 30, and the MiG-35", Quora, https://www.quora.com/How-does-the-

iv Saab-Gripen-E-compare-to-the-Su-35-Su-30-and-the-MiG-35

v   "How does the J-10 compare with the F-16, Gripen, and the F-2? All in their most modern configurations",

vi https://www.quora.com/How-does-the-J-10-compare-with-the-F-16-Gripen-and-the-F-2-All-in-their-most-modern-configurations

vii Jörgen Furuhjelm, Johan Segertoft, Joe Justice, and J.J. Sutherland, "Owning the Sky with Agile, Building a Jet Fighter Faster, Cheaper, Better with Scrum," https://www.Scruminc.com/wp-content/uploads/2015/09/Release-version_Owning-the-Sky- with-Agile.pdf

viii Oxygen generator system from Honeywell. https://aerocontent.honeywell.com/aero/common/documents/myaerospacecatalog-documents/Defense_Brochures-documents/Life_Support_Systems.pdf

ix Martin-Baker seat, MK10 Ejection Seat. http://martin-baker.com/products/mk10-ejection-seat/

x   Steubkeller S., et al. "Modeling and Simulation of Saab Gripen's Vehicle Systems, Challenges in Processes and Data Uncertainties," 27th International Congress of the Aeronautical Sciences, ICAS 2010, Nice Fr., Sept 2010.

A Student Team Designs a Prosthetic Arm Using Scrum Methods

A CASE STUDY FOR THE MECHANICAL DESIGN PROCESS

INTRODUCTION

A team of four students at Olin College created a below-the-elbow prosthesis for amputees using the Agile/Scrum design process. This team spent eight weeks designing the "Smart Arm" complete with an intuitive feedback controller using commonly available DIY components.

The Smart Arm was designed Victoria, Liani, Celine and Ellie, students in a mechatronics class taught by Professor Aaron Hoover. He included the Scrum framework so the students could experience the process while they designed a device requiring mechanical, electronic, software and systems engineering skills. Prof. Hoover felt that the Scrum methodology well supported interdisciplinary teams and would increase their chances of design success. The students could design any product they wanted that had mechanical, electrical/ sensing and real-time control components. The eight-week project was structured so that there were four two-week sprints. At the end of the project the team had to produce a demonstratable prototype. The Smart Arm team's result is shown with the outer covering off on the left next to human arm in Fig. 1.

Figure 1. The partially assembled Smart Arm next to a human arm

- **The Problem:** Learn mechatronics by developing a device that emphasizes each student's individualized learning objectives and lets all experience an engineering design best practice.
- **The Method**: Use the Scrum process to manage product evolution and team communication increasing the chances of success.
- **Advantages**: Scrum allowed the team to go from zero to a working am in eight weeks enhancing mechatronic design, communication, and decision-making skills.

BACKGROUND

The Smart Arm project was part of an experiential class on mechatronics taught by Prof Aaron Hoover of Olin College. The learning objectives for this course were, in Prof Hoover's words:

"At the end of this course, students will be able to:
- Work effectively as a member of a project team.

- Develop design concepts and create technical specifications that address defined needs.
- Balance trade-offs and make defensible choices among design alternatives.
- Use modern tools to construct mechatronic systems.
- Assess and select appropriate components for mechatronic circuits and systems.
- Use written, oral, and graphical communication to convey design ideas and solutions, electronic system analyses, and experimental results.
- Undertake an iterative prototyping process to improve design ideas."

To meet these objectives, early in the term he introduced the Scrum design process. His goal in using Scrum was to provide the students with a framework giving the multidisciplinary teams a high chance of design success.

After Prof. Hoover introduced mechatronic basics during the first few weeks of the term, the class self-chose four-person teams and picked their own projects that had to include:

- A non-trivial mechanical system
- A non-trivial electrical/sensing system
- Real-time control (i.e., must use a microcontroller)

The teams had eight weeks to develop their products and were to work in two-week sprints. His grading system gave 65% of the course credit for the deliverables presented during the Sprint Review at the end of each sprint. He introduced the Scrum framework with a slide presentation built around Fig. 2.

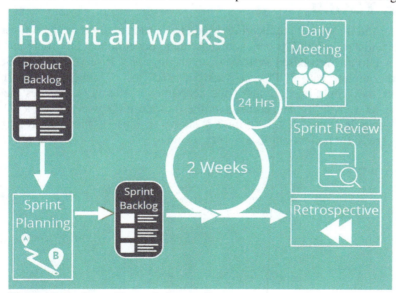

Figure 2. Professor Hoover's Scrum diagram.

Professor Hoover leads about 20 teams a year through this class. The projects cover a wide range: e.g., games, music generators, wearables, cooking devices, bike products, visual arts, and prosthetics. The team featured in this case study consisted of Victoria, who focused on the controller and software; Celine who identified and implemented the sensors; and Liani and Ellie, both mechanical engineers, who developed the physical components and their manufacture. All were sophomores except Victoria, a junior.

At the beginning of the project, they had to "pitch" their idea about what exactly to design to each other and then to Prof. Hoover. No one remembers who first proposed the Smart Arm, but all were all drawn to the idea that "hobby" technologies such as 3D printing, Arduinos, and hobby servos can be used to improve day-to-day experiences for prosthetic users. They also liked that this project would allow each of them to learn more about a skill they were eager to practice: e.g.,sensor-feedback design, simple state-machines, design for manufacture.

To guide the students, the class was formatted in 2-week sprints with a certain amount of progress expected in all subsystems (electrical, software and hardware) at the end of each sprint. Further, they were judged on how well they identified and managed risk, how they made decisions, and how clearly they articulated the next sprint's goals.

Each team was given a budget of $250 covered by Olin, but they could spend more using personal money if they wanted. The Smart Arm team spent $214.

THE TOOLS

Besides the traditional engineering design tools such as solid modeling, spreadsheets and Arduino IDE (open- source Arduino development environment) they used Trello[1], a web-based system for managing lists. It served as their Scrum Board complete with a "Project Backlog," "To Do," "Doing," as seen in Fig. 3. Here they could post the tasks and keep track of the work. Besides the lists shown, as each sprint was completed, the finished tasks were put on a "Done" list. A sample Done list for Sprint 3 is shown in Fig. 11. The team found that using Trello for collaboration was very useful.

---

1   www.trello.com

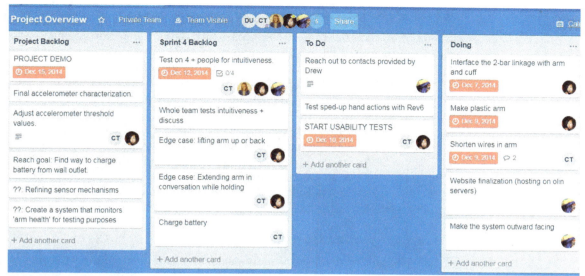

Figure 3. Partial Trello Scrum board

Additionally, they logged their progress in blog posts. See a typical example in Fig 4. Creating a project website or blog was part of the course requirements. During the project, there were 36 blog posts, most with images or videos. All the team members posted their progress, and this became their repository for product information. Where Trello was an inward facing communication tool, the blog was outward facing.

Figure 4. Sample blog post

THE PROCESS

Rather than track the project chronologically through the four sprints, examples from them will be used to explore what the team did well and where the project was lacking. This retrospective is by no means complete as the team's Trello cards, their blog, and other material captured a rich history of the project. Only examples sufficient to support the important Scrum learning objects are included.

The process followed by the team closely adhered to that shown in Fig. 2. It must be remembered in reading this material that the high-level goals for the project were to learn about mechatronics with each individual given the opportunity to learn more about their technical area of interest. The use of the Scrum framework was to both facilitate the mechatronic learning and to experience a design process best practice.

Team Organization

Usually, Scrum teams are 4-9 individuals with one person filling the role of Scrum Master – driving the process so the technical team can operate efficiently, a second person being the Product Owner – representing the voice of the customer and the remainder being the technical team. Since this team was only four people in an academic situation, the team structure was simplified. Prof. Hoover assumed the role of Product Owner since the team's sprint reviews (a formal chance to get feedback on their work) served as the graded deliverable capping each sprint. There was no Scrum Master on this team, but some teams choose to have one person fill that role while also being part of the technical team.

Figure 5 Two team members working together.

Further, in the ideal software Scrum team, the members have broad enough knowledge and experience to do many of the needed tasks. Here, one goal was for each student to gain knowledge and experience in their individual fields. So, the team broke into "expertise" areas, corresponding with individual learning goals. Luckily the areas of interest were independent enough that allocation of tasks was not an issue. For tasks that didn't line up with a specific area of expertise or learning objective, someone would volunteer during team meetings.

Additionally, as in many multidisciplinary software/hardware systems, there was little overlap in expertise, and thus tasks were often completed by the only person. That is not to say that the team did not work together. Collaborative efforts were common as in Fig 5.

Design Goal Development

Since the class was not primarily focused on making a product, the amount of domain research and goal development was somewhat limited. In retrospect, team members felt that they should have more spent more time up-front understanding the users and the issues.

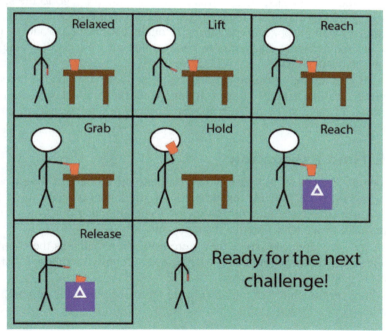

Figure 6. The system states for the Smart Arm

The team did not write formal user stories or develop customer requirements. However, partway through the effort, they realized that they did not have a clear picture of exactly what actions the Smart Arm needed to do and developed the system state model in Fig. 6. Here the steps that the hardware and software needed to accomplish were codified. This diagram helped focus the team for the remainder of the project.

Product Backlog Management

The team's use of Trello (Fig. 3) made backlog management easy. As new requirements were realized they were added to the Product Backlog list. While there was no formal rank ordering in this list, the team paid attention to sequencing - those items that would delay someone else were dragged up nearer the top of the list. They were driven primarily by the Scrum cadence of a sprint review every two weeks.

Task Identification, Measures, Targets and Tests

The team did not follow any formal method to generate "tasks." The titles on the Trello cards were generally "actions" to be taken (see Figs 3 and 11) with no formally stated measures, targets, and tests.

For some of the tasks, "done" was clear. For example, "go to the store to pick up x, y, z" having an obvious conclusion. For other tasks, the deliverables were driven by an unspoken expectation for quality of finished products. Often this would require team discussion to confirm that all were on the same page about the status of a task.

There was implicit "testing" to address those more open-ended tasks. For example, a task like "verify sensor filtering" would include a test in which the sensors would have to be hooked up, run through several scenarios, and the results discussed.

Task Time Estimation

Each person "assigned" to a task would estimate the amount of time it would take. Some drew on previous experiences on similar tasks, some used timeboxing force a projected amount of time on something, and others would do a little research to see what external resources might help.

For the software tasks, since they were using an Arduino as the controller, the collective experiences of the open-source community members for similar projects helped project timeboxing.

In later sprints, time estimation became more accurate as is common with Scrum teams.

Sprint Focus

The team had long discussions over which tasks to work on during each sprint. The team balanced the product need with their specific learning goals to allocate items to each person. They were also driven by "It would be awesome if we showed this off at the sprint review."

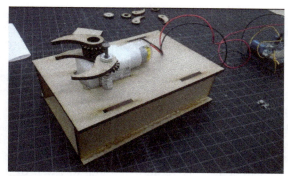

Figure 7. First prototype

During team meetings, they added to the lists in Trello and reordered the cards to help manage the sprint focus.

Risk and Uncertainty Management

This exercise was all about learning; learning about the technologies and learning about Scrum. Learning removes uncertainty, as can clearly be seen by the team's progress through their prototypes and models (Figs 7, 8, 9, and 14).

When they began, they were uncertain about how to grip objects, how to sense the gripping force and how to provide feedback to the wearer. The first model, built very early in Sprint 1, consisted of a laser-cut wooden box that was hot glued together (Fig. 7). It had simple pincers geared together as a gripper. A potentiometer was rigged to give feedback to control the gripping. It could respond to sensor input and gripped or released based on system state. It only took about 2.5 hours to build, but the base was not sturdy, and the gear drive slipped due to poor tolerance and uncertain placement. Also, the team had set a goal of being able to pick up a cup, and the fingers proved too short to accomplish this task. Despite these limitations, this simple model taught the team a lot about gripping, and sensing.

Figure 8. Second prototype in Sprint 1

The second model (Fig. 8) was built later in the first sprint and was part of the "show and tell" for the first Sprint Review. It used the same sensors and inputs as the first model but had more sophisticated grabbing action and a breadboarded controller circuit to manage sensing.

As part of this the first sprint, one of the team members began to research sensing. They wanted to sense both the gripping force and how they might control it through arm and elbow movement.

The Smart Arm is a "below-the-elbow" prosthetic, which assumes that the elbow joint is intact and able to be actuated. Through sensing the elbow position or motions, the team designed the controller to "intuit" the user's intent. For example – a user extending their arm, as though to reach or grab something - could implicitly tell the hand to close. Alternatively, if the hand was already holding something, extending the arm could imply that the user would like to release.

To sense the user's intent, they began by considering piezoelectric, electrode muscle sensors, flex sensors, and stretch sensors. They then found some very inexpensive flexible customizable sensors at Adafruit, a DIY supplier[2].

Based on what was learned about gripping and sensing in Sprint 1, Sprint 2 began with work on "Arm 1.0".

Made of cardboard, eyehooks, blue foam, spring wire, and thread (Fig. 9), Arm 1.0 was a great thought experiment around arm size, actuation technique, and aesthetics. After building it, they began moving into creating a real-life model that integrated the servo and motor systems with 3D printed fingers, and the sensors.

Known Stable Interface Definition

One strong design feature in Scrum for hardware is to design known stable interfaces early in the process. The team identified interfaces:
- Between the human arm and the Smart Arm
- Between the physical structure and the PC board
- Between the hand and arm (Seen as the joint between the blue foam hand and cardboard arm in Fig.9.)

The team made an effort to define these interfaces early in the project and keep them fixed.

---

2 DIY Sensor film kit. http://www.adafruit.com/product/1917

Figure 9. Arm 1.0 built early in sprint 2.

Modular Design Creation

With the interfaces determined work could progress on the modules. The arm structure (for example) rapidly progressed from the cardboard in Fig. 9, through the laser-cut wooden model in Fig. 10. Here the wooden bulkheads of the arm as connected with threaded rods. This module later was transformed into a plastic structure as seen in Figs. 1 and 14.

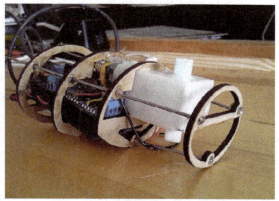

Figure 10. The laser-cut wooden arm with early PC board mounted.

Sprint Standup Meetings

The team did "stand-ups" at every all-hands meeting. These occurred in class twice a week and additionally 1- 3 times a week outside of class. Team members also met as sub-teams (2-3 members at a time) 1-3 times a week. Each team member put in a minimum of 10 hrs./week outside of class time, though most weeks it was quite a bit more (20+ hrs./week/person). Seventy percent of their time was spent in groups of two or more. While each was enrolled in at least three other classes at the time, they all wished that this was their only class so that they could spend more time on the project. This feeling is common in project design classes.

The team did "stand-up" type communication at the beginning of every meeting. At the end of a meeting, they would also re-iterate everyone's task "doing" items, so they would each know what was expected of them by the next meeting. Between meetings, they used instant messaging or impromptu chats in the dining hall and dorms to keep up.

Sprint Progress Tracking

The team used Trello to track sprint progress. As seen in Fig 11, an example of "Done" tasks from Sprint 3, each of the Trello cards provided a history of work accomplished.

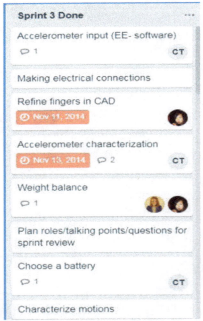

Figure 11. Trello list showing what was completed in Sprint 3.

Sprint Review

At the end of each sprint, the team presented their deliverables as part of the Sprint Review.

Sprint reviews involved presenting progress, in the form of power point presentations or demonstrations, to the class at large. It involved discussing progress, hold-ups/sticking points, and intents for the next sprint cycle. This was an important forum to receive feedback from voices outside the team on project pacing, prioritization of various goals, and design.

Sprint Retrospective

The team used Google Sheets to manage team retrospectives (Fig 12, a sample from the first sprint). The use of a spreadsheet allowed them to do some self-reflection and posting of thoughts individually before the retrospective meeting.

They set up four columns on the spreadsheet to capture which sprint (SCRUM), what went well (PLUS), what they would change (DELTA) and what to do about it (SOLUTION). Further, they each entered their thoughts using a different color.

It is worth noting that even in Sprint 1, there was the realization that tasks were not well defined (2nd from the bottom DELTA entry).

SCRUM	PLUS	DELTA	SOLUTION
1	Whole-team understanding of system components	May have to step away from this as things get more in-depth/specialized?	Good documentation!
1	Dropbox for easy access to subsystems	I need to refresh my stuff in there more often	
1	We're all civil to each other	Have a clearer idea of what we want to explore in terms of teaming per scrum	Revisit teaming conversation, chat with Aaron about talking with us for a longer amount of time
1	People following through on deliverables	Updating when a deliverable isn't likely to be met before a deadline	Quick email updates; remember to also tell people about falling behind
1	Fast turn around time, with system integration in mind	Sharing goals for a design (software or hardware) before implementation	Delta <- Failure Mode
1		Change work environments (i.e. work in hallway on occasion)? May / may not be feasible; I find that changing enviirons helps clear my mindset sometimes	
1	Using trello to kick-off meetings	Not using trello very dynamically (maybe ok)	Make more concerted efforts to make quick checks on trello during small meetings
1	Emphasis on documentation and thoroughness	Updating in different places (PDM, Drive, Trello, Email)	Update website and jazz; blog section
	Fast prototyping	Tasks not fully described	Endgoals on trello for each [sub]system
	Just-do-it mindset	License to do things individually? vs. in subteams	Do stuff! Just inform all!

Figure 12. Sample retrospective entries

WHAT WAS LEARNED

The final Smart Arm is shown in Fig. 13 with one of the team members simulating its use and in Fig. 14 with its outer cover off. The resulting product is quite an accomplishment for eight weeks of work while taking other classes. The team felt that they learned a lot and were all quite happy with their results.

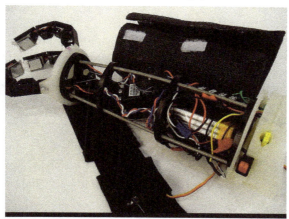

Figure 13. The Smart Arm in simulated motion

In terms of the course's learning objectives, Scrum directly enabled the students to:

- Work effectively as a member of a project team by scaffolding the distribution of work across team members, project time, and project objectives.
- Develop design concepts and create technical specifications that address defined needs.
- Undertake an iterative prototyping process to improve design ideas.

The Scrum process worked very well for a multidisciplinary team learning about mechatronics. Based on this success, it is clear that the framework can be used in other design experience courses. It is also clear that the structured form of meetings in Scrum greatly aided in communication both internally and with customers (i.e., the professor and classmates).

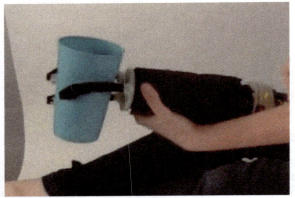

Figure 13. The final Smart Arm with the covering removed.

What was accomplished regarding the Scrum framework is summed up in Table 1.

Table 1 Summary of Scrum learning

Scrum learning objective	Result
Team Organization:	The agile/ Scrum methodology generally fostered student ownership of the product and process
Design Goal Development:	Design goals were poorly developed. In retrospect, one team member said: "I believe that spending more time in the ideation/research phase would have improved this project a lot."
Product Backlog Management:	The use of Trello worked very well for this team.

Table 1 *continued*

Scrum Learning Objective	Result
Task Identification, Measures, Targets and Tests:	The Trello cards were task-centric. However, functional targets were not set with clear measures, targets, and tests. The main driving force the desire to have something to demonstrate at the next sprint review. Prof. Hoover pushed each team for a revised definition of their "minimum viable product" at the end of every sprint based on what they have learned about their problem in the last two weeks.
Task Time Estimation:	Efforts were made to tailor task lengths to sprint schedule although this was done informally with no recording of estimates.
Sprint Focus:	Good use of Trello to manage the Sprint Backlog. A method for choosing which tasks to tackle was ad-hoc.
Risk and Uncertainty Management:	This project was all about learning, the reduction in uncertainty. While no formal methods to address risk were used, the students used the early sprints to learn, iterate and zero in on a working product.
Known Stable Interface Definition:	While not a conscious effort, stable interfaces were established between the user and the prosthesis, and between the frame and the control boards, and between the arm and the hand.
Modular Design Creation:	Not a large effort here, but the interfaces would allow multiple boards and hands very easily.
Daily Meetings:	The team did an excellent job of frequent meetings. Being in an academic situation fairly much precluded daily meetings at a set time.
Sprint Progress Tracking:	The Trello lists were well used to track sprint progress. No burndown chart was used.
Sprint Review:	The structure of the course included a sprint review every two weeks.
Sprint Retrospective:	The retrospective notes kept in Google Charts was well used.

Both Prof Hoover and his students at Olin should be proud of this effort. The students enjoyed the experience and even a couple of years removed from it thought it a great learning experience.

ACKNOWLEDGMENTS

This case study was supported by Prof Aaron Hoover of Olin college and his students: Eleanor Funkhouser (class of 2017), Liani Lye (class of 2017), Victoria Preston (class of 2016), and Celine Ta (class of 2017). They kindly shared their blog, Trello lists, and other material. Further, they answered detailed questions about their experience, and much of the material here is paraphrased from their responses.

CPSIA information can be obtained
at www.ICGtesting.com
Printed in the USA
BVHW011633150421
605033BV00006B/367